Palgrave Studies in Disaster Anthropology

Series Editors
Pamela J. Stewart
Department of Anthropology
University of Pittsburgh
Pittsburgh, PA, USA

Andrew J. Strathern
Department of Anthropology
University of Pittsburgh
Pittsburgh, PA, USA

This book series addresses a timely and significant set of issues emergent from the study of Environmental [sometimes referred to as "natural"] disasters and the Series will also embrace works on Human-produced disasters (including both environmental and social impacts, e.g., migrations and displacements of humans). Topics such as climate change; social conflicts that result from forced re-settlement processes eventuating from environmental alterations, e.g., desertification shoreline loss, sinking islands, rising seas.

More information about this series at
http://www.palgrave.com/gp/series/15359

Lyn Carter

Indigenous Pacific Approaches to Climate Change

Aotearoa/New Zealand

palgrave
macmillan

Lyn Carter
University of Otago
Dunedin, New Zealand

Palgrave Studies in Disaster Anthropology
ISBN 978-3-319-96438-6   ISBN 978-3-319-96439-3   (eBook)
https://doi.org/10.1007/978-3-319-96439-3

Library of Congress Control Number: 2018949927

This Palgrave Pivot imprint is published by the registered company Springer Nature Switzerland AG
The registered company address is: Gewerbestrasse 11, 6330 Cham, Switzerland

SERIES EDITORS' PREFACE

KNOWLEDGE ENCOUNTERS: PLANNING FOR CLIMATE CHANGE IN AOTEAROA/NEW ZEALAND

Lyn Carter's study, as she explains, is a companion and a counterpart to Jenny Bryant-Tokalau's study of environmental issues and adaptation among Pacific Islands peoples. The overall messages of these two books dovetail clearly: indigenous environmental knowledge is an important resource in plans and practices for both mitigation of the effects of environmental changes and adaptations to these changes by innovative future-oriented strategies. Dr Carter's study, in this book, presents us with a very special and instructive insight into this complex arena of problems, by focusing on her own Māori group, Kai Tahu Iwi, in the South Island of New Zealand. It also deals continuously and seriously with how Māori/Iwi and their TEK (Traditional Ecological Knowledge) are engaged with national-level government planning for environmental management, in a context of looming and actual climate change. The ideal of combining, rather than opposing, Māori TEK and what is labelled as scientific knowledge is explored productively in this book. The essential point is that TEK and its creative applications can add valuably to efforts to handle the social impacts of climate change.

Dr Carter's forward-looking and inspiring approach links together worlds that are more often kept separate. She begins with Māori myths, or origin stories, about the world, especially those relating to the tension between the environmental deities of the land and the sea. Māori groups' stories also link them to the broader Pacific islands, areas from which their ancestors came, to inhabit what their famous explorer Kupe is said to have called Aotearoa, land of the long white cloud. We ourselves have visited the beach area in Rarotonga, in the Cook Islands, held locally to be the launching place for the first canoes that sailed to Aotearoa. We have watched how the ocean waves beat on the reef and how there is a favorable passage out from the bay via a gap in the reef. A feature of Dr Carter's book is that she acknowledges and brings to life this connection with Polynesian seafaring history.

Māori see themselves as belonging equally to land and sea and the struggle for dominance of the *atua* or deities of these two parts of the lifeworld. Her account always starts from some aspect of TEK, and then moves to examine the engagement of Māori tribespeople (*iwi*) with the New Zealand government and its own attempts to develop an environmentally sensitive program for the use of resources. Central to this process is the need to recognise Māori *mahika kai*, places that are resources for food, and how to maintain these without environmental degradation. One of the many interesting parts of the book deals with Kati Huirapa hapu and their efforts to restore and maintain wetlands around the estuaries of the Waikōuaiti and Waihemo rivers that are catchments for the *inaka* (whitebait) fish. The Kāti Huirapa name for the wider catchment area involved is Matainaka, pointing to the importance of the whitebait as a food source. Whitebait are freshwater fish, and salt water encroachment and floods place their habitat at risk, so Kāti Huirapa have researched the most appropriate grasses to grow at river edges that give stronger protection for the inaka spawning sites than the native grasses which they were previously replanting there.

Dr Carter's comments here are twofold. One is to point out the significance of place, or more precisely emplacement. As our author aptly notes: "The name [Matainaka] recalls the place, the place recalls the stories, the stories recall the whakapapa [genealogies]; and from that each group knows how they belong and fit within the environment". Her formulation here applies, beyond the immediate context, on a pan-Pacific basis, and emphasises why place names are so evocatively significant in projects of environmental restoration and conservation.

Other arenas in which Māori groups are closely involved, along with nongovernmental organisations (NGOs) and government initiatives, are sand-dune reclamation and protection against erosion, and also blue carbon sink creation in which mangrove swamps, beds of sea grasses, and tidal marshes in their pristine forms act as traps for carbon and help to reduce greenhouse gas emission. When wetlands are drained or damaged, the carbon is released into the atmosphere. Māori stewardship of such coastal areas is therefore vital. Issues that remain include those that have to do with the environmental impact of dairy agriculture, and of commercial logging activities that reduce forest cover. These issues go beyond TEK and its Māori version MEK (Māori Ecological Knowledge), but there is a special role for Māori values in the coastal contexts where much Māori settlement of the South Island took place. Throughout, Dr Carter insists both on the importance of traditional knowledge and on its dynamic capacities for change. In general, then, her book is a powerful exemplification of the significance of indigenous ideas in the sphere of environmental studies. The first Māori explorers would have brought with them much IK (Indigenous Knowledge) from the earlier settlement of the Pacific Islands and their ancestral progenitors there; but the interest of the Māori case, especially of Kai Tahu in the South Island (Te Wai *Pounamu* in Māori terms) lies in how they adapted their IK to new ecological circumstances, for example by extraction of sugar from the Ti tree and the creation of storage pits for sweet potatoes.

Dr Carter not only introduces us to the world of Māori knowledge but also painstakingly takes us through Māori encounters with government legislation, the work of NGOs, and international concerns in the field of environmental protection and renewal. Her book, along with the companion book in this same series by Dr Bryant-Tokalau on Fiji and other regional island case studies, puts Pacific knowledge systems and practices firmly in the contemporary intellectual picture as embodying creative efforts to deal with the existing and likely future impacts of climate change in the world at large (other case studies in the Pacific have foregrounded this perspective: see Stewart and Strathern, 2015 and 2018; Strathern, Stewart, Carucci, Poyer, Feinberg, Macperhson, 2017).

Pittsburgh, PA, USA   Pamela J. Stewart (Strathern)
                                                                  Andrew J. Strathern

REFERENCES

Stewart, P. J., & Strathern, A. (Ed.). (2015). Disaster Anthropology. In *Research Companion to Anthropology* (pp. 411–422). Oxford: Routledge.

Stewart, P. J., & Strathern, A. (2018). *Diaspora, Disasters, and the Cosmos: Rituals and Images*. Durham: Carolina Academic Press.

Strathern, A., Stewart, P. J., Carucci, L. M., Poyer, L., Feinberg, R., & Macpherson, C. (2017). *Oceania: An Introduction to the Cultures and Identities of Pacific Islanders* (Updated and Revised, 2nd ed.). Durham: Carolina Academic Press.

ACKNOWLEDGEMENTS

The ideas for this book came from conversations with my University of Otago colleague and Pacific researcher, Jenny Bryant-Tokalau in 2015. Discussions about the many ways Pacific peoples have been finding indigenous solutions to environmental disasters, including the impact of climate change, were at the forefront of the conversations and subsequent research. It became apparent that here in Aotearoa/New Zealand (A/NZ) there is a dearth of adaptation strategies and practices happening (indigenous knowledge informed or otherwise). During the course of the conversations, two books emerged: one that examines the Pacific Island Countries' (PIC) responses to climate change; and another one that examines A/NZ's lack of response and what indeed they could learn from their Pacific neighbours. In short, A/NZ needs to take note and heed the lessons from the Pacific.

My research and writing has been focused on areas that concern (mainly) my own Māori tribal group (Iwi), Kāi Tahu. More importantly the ecological knowledge discussed in some examples refers to the region where I live and experience the knowledge and practices first-hand. With the identified threats from sea-level rise, flooding, and wildfires come potential changes to the way we currently experience our cultural landscapes and environment. The challenges facing our future generations will demand flexible decision-making in how to adjust and plan for those challenges. These may include relocations and future limits to accessing our mahika kai (resources), many of which are part of the coastal and wetland ecologies that fringe our tribal territories. The present generations are looking at ways to lessen the impact on our vulnerable resources to ensure future access and use. The knowledge added to the many ways of knowing how to work with our environment will help alleviate any cultural and environmental tipping points that may affect the mahika kai. The many stories within the book come from research into the past, conversations with elders in the present, and offer dreams and aspirations to take this wisdom into the future. I therefore acknowledge those Kāi Tahu who came before, those who are still living, and those who will come later into a world much changed through the impact of climate change. I acknowledge too the many colleagues both within A/NZ and internationally who have contributed their thoughts and research to my own, which has enabled the continuing conversation in indigenous knowledge frameworks as relevant ways of understanding and operating in the world.

Kā mihi mahana atu ki a koutou, kā rangatira kairangahou i ēnei mātauranga, i ēnei whawhai.

Tēnā koutou katoa.

CONTENTS

1 Introduction 1

2 Setting the Scene 15

3 Traditional Ecological Knowledge in Climate Change 25

4 Aotearoa/New Zealand and Land-Use Changes 39

5 Aotearoa/New Zealand and the Emissions Trading Scheme 55

6 Aotearoa/New Zealand Adaptation Strategies and Practices 71

7 Where to from Here? Learning from Our Pacific
Neighbours 85

References 97

Index 103

ABOUT THE SERIES EDITORS

Pamela J. Stewart (Strathern) and Andrew J. Strathern are a wife-and-husband research team who are based in the Department of Anthropology, University of Pittsburgh, and co-direct the Cromie Burn Research Unit. They are frequently invited as international lecturers and have worked with a number of museums to assist them in documenting their collections. Their published work includes over 50 books and over 250 articles, book chapters, and essays on their research in the Pacific (mainly Papua New Guinea and the South West Pacific region, e.g. Samoa and Fiji); Asia (mainly Taiwan, and also including Mainland China and Japan); and Europe (primarily Scotland, Ireland, and the European Union countries in general); and also New Zealand and Australia. Their most recent co-authored books include *Witchcraft, Sorcery, Rumors, and Gossip* (2004); *Kinship in Action: Self and Group* (2011); *Peace-Making and the Imagination: Papua New Guinea Perspectives* (2011); *Ritual: Key Concepts in Religion* (2014); *Working in the Field: Anthropological Experiences Across the World* (Palgrave Macmillan, 2014), and *Breaking the Frames: Anthropological Conundrums* (Palgrave Macmillan, 2017). Their recent co-edited books include *Research Companion to Anthropology* (2015); *Exchange and Sacrifice* (2008) and *Religious and Ritual Change: Cosmologies and Histories* (2009 and the updated and revised Chinese version: 2010). Stewart and Strathern's current research includes the topics of cosmological landscapes; ritual studies; political peace-making; comparative anthropological studies of disasters and climatic change; language, culture and cognitive science; and Scottish and Irish studies. For many years they served as Associate Editor and General Editor (respectively) for the *Association for Social Anthropology in Oceania* book series and they are co-series editors for the *Anthropology and Cultural History in Asia and the Indo-Pacific* book series. They also co-edit five book series: *Ritual Studies*; *Medical Anthropology*; *European Anthropology*; and *Disaster Anthropology* and *Anthropology and Cultural History in Asia and the Indo-Pacific* and they are the long-standing co-editors of the *Journal of Ritual Studies* [Facebook: https://www.facebook.com/ritualstudies]. Their webpages, listing publications, and other scholarly activities are:

http://www.pitt.edu/~strather/ and
http://www.StewartStrathern.pitt.edu/

GLOSSARY OF MĀORI LANGUAGE
AND TERMINOLOGY

Note: Kāi Tahu dialect is utilised throughout this book. Exceptions are where standard te reo Māori (Māori language) has been used in direct quotes and therefore remains unchanged.

Ahi kā	Warm fires (occupation)
Atua	Ancestors, supreme beings
Hapū	Sub-tribe, pregnant
Ira atua	Spiritual elements of all living things
Ira takata	Physical/pragmatic elements of all living things
Iwi	Tribe
Kāi Tahu	Prominent South Island Māori tribal group
Kaitiakitaka	Guardianship, care, responsibility
Ki uta ki tai	From the mountains to the sea
Mahika kai	Resources; resource-gathering sites
Mana whenua	Power and authority
Māori	Indigenous peoples of Aotearoa/New Zealand
Mātauraka Māori	Māori knowledge frameworks
Mātauraka-a-Iwi	Tribal knowledge frameworks
Papatūanuku	Earth
Rakatirataka	Leadership, management
Rakinui	Sky and heavens
Rūnaka	Multi-hapū regional councils (as part of the Kāi Tahu tribal governance and management structure)
Takaroa	Māori supreme ancestor of sea and all the living things within it
Takiwā	Tribal territories

Tāne	Māori supreme ancestor of forests and birds
Taoka	Treasured and culturally significant objects and people
Tāwhirimatea	Māori supreme ancestor of winds and weather patterns
Te Ika a Māui	The great fish of Māui/North Island of New Zealand
Te Wai Pounamu	The Greenstone land/South Island of New Zealand
Tikaka	Tribal cultural practices
Tūrakawaewae	Standing place for the feet
Wāhi taoka	Places of special cultural significance
Wāhi tapu	Places of high spiritual value
Whakapapa	Genealogies, layers of relationships
Whakatauki	Māori sayings
Whakawhanukataka	Managing relationships
Whānau	Family group
Whenua	Land, placenta

Acronyms

AAU	Assigned Amount Units
ADB	Asian Development Bank
A/NZ	Aotearoa/New Zealand
AOSIS	Alliance of Small Island States
CCS	Carbon Capture and Storage
CDM	Clean Development Mechanism
CERs	Certified Emissions Reduction Units
CITES	Convention on International Trade in Endangered Species of Wild Fauna and Flora
CS	Continental Shelf
CSCC	Comprehensive Strategy on Climate Change
DCC	Dunedin City Council
EEZ	Exclusive Economic Zone
EMRs	Emissions Reduction Units
EPA	Environmental Protection Agency
ESCAP	The United Nations Economic and Social Commission for Asia and the Pacific
ETS	Emissions Trading Scheme
GEF	Global Environment Facility
GHE	Greenhouse Effect
GHG	Greenhouse Gasses
IL	Indigenous Knowledge
IPCC	Intergovernmental Panel on Climate Change
IPMPCC	Indigenous People, Marginalised Populations and Climate Change
JI	Joint Implementation
KHR	Kāti Huirapa Rūnaka ki Puketeraki
LA	Local Authorities

LG	Local Government
LGNZ	Local Government New Zealand
LIM	Land Information Memorandum
LTCCP	Long-Term Council Community Plans
MAB	Man and the Biosphere Program (UNESCO)
MEK	Māori Environmental Knowledge
MfE	Ministry for the Environment
NGOs	Nongovernment Organisations
NIWA	National Institute of Water and Atmospheric Research
NZETS	New Zealand Emissions Trading Scheme
NZU	New Zealand Government Units
OGEEI	Oil and Gas Exploration and Extractive Industries
PIC	Pacific island Countries
PIDF	Pacific Island Development Forum
RMA	Resource Management Act
RMUs	Removal Units
RoO	Te Rūnaka o Ōtākou
SAM	Social Accounting Matrix
SCBD	Secretariat of the Convention on Biological Diversity
SGG	Synthetic Greenhouse Gas
SME	Small to Medium Enterprises
TEK	Traditional Environmental Knowledge
TRoNT	Te Rūnanga o Ngāi Tahu is the Ngāi Tahu tribal structure for the economic, cultural, social, and environmental development. The governance board is made up of one representative from each of the 18 Kāi Tahu regional councils
UNDP	United Nations Development Programme
UNESCO	United Nations Educational, Scientific and Cultural Organization
UNFCCC	United Nations Framework Convention on Climate Change

LIST OF FIGURES

Fig. 1.1 Map of Oceania, *Central Intelligence Agency*, 2010. The
         'Polynesian Triangle' comprises Aotearoa/New Zealand (in the
         south), across to Rapanui (Easter Island, in the east), and the
         third apex in the north, the Hawaiian Islands   11

Fig. 3.1 Photo of Takaroa. Carved pou (post) at Warrington Beach,
         Dunedin. (Carved by members of Te Whare Wananga o Te
         Whānau Arohanui, Watiati, Dunedin. The spirals represent the
         past, present, and future generations of kaitiaki for the resources
         around the East Otago coastline—the domain of Takaroa. It
         stands at the interface between sea and land)   30

Fig. 6.1 Photo of DCC notification of pikao protection at Brighton
         Beach, 2015   80

LIST OF TABLES

Table 5.1   New Zealand's GHG units   60
Table 6.1   Principal determinants of Māori community sensitivity and
            adaptive capacity   74

CHAPTER 1

Introduction

Abstract The book *Indigenous Knowledge and Climate Change: Aotearoa/ New Zealand* has its genesis in conversations with my colleague, Jenny Bryant-Tokalau on how Aotearoa/New Zealand could benefit from many of the Pacific ways of understanding and dealing with environmental change across the Pacific. Aotearoa/New Zealand as a Pacific Island nation has much to learn from indigenous ways of knowing and understanding mitigation and adaptation brought about through the impacts of climate change. This chapter introduces how Aotearoa/New Zealand can benefit and learn from its Pacific Island neighbours and key to this is utilising Māori knowledge frameworks and practices. From an indigenous knowledge perspective, relationships between people and the other elements of an ecosystem are dynamic and constantly changing, thus requiring renegotiation to overcome challenges that present themselves.

Keywords Pacific Island countries • Climate change • Indigenous knowledge • Aotearoa/New Zealand

The 2014 Intergovernmental Panel on Climate Change (IPCC) report emphasised that Aotearoa/New Zealand (A/NZ), along with smaller Pacific Island countries (PIC), is unavoidably exposed to the effects of climate change with key risk areas of sea-level rise, flooding, and wildfires.[1]

© The Author(s) 2019
L. Carter, *Indigenous Pacific Approaches to Climate Change*,
Palgrave Studies in Disaster Anthropology,
https://doi.org/10.1007/978-3-319-96439-3_1

Climate change is undeniably among us and we cannot be unaware of the impacts on our wider Pacific neighbours. The New Zealand government has been accused of 'lethargy' and the government's policies exhibiting 'an indifference to the phenomenon of climate change both at international level and domestically'.[2] A/NZ as a large industrialised nation amongst her Pacific neighbours has to mount a two-pronged defence against climate change—both mitigation against increased greenhouse effect (GHE), and secondly developing adaptation strategies for the impacts from climate change already being felt. The IPCC has called for a combination of the two strategies to achieve the most positive benefits.[3] With a primary focus on mitigation to date, A/NZ has paid scant attention to adaptation factors necessary for dealing with the many challenges and disasters that will come with the changing climate. Despite being part of Oceanic Polynesia, 13 per cent of New Zealanders are sceptical that anthropogenic climate change exists.[4] Of the 15 industrialised countries monitored, only Australia and Norway rate higher with the United States rating just below New Zealand. The reasons for this scepticism are complex, but according to a recent report from Tranter and Booth, the key consistent factors are affiliation with conservative political parties, gender, being unconcerned about the environment, having little trust in government, and a correlation with CO_2 emissions.[5] Tranter and Booth observed those who favour economic growth above environmental interests are 'those who … tend to believe that global climate change is not occurring, that the causes of global climate change are more natural than human caused, and that its consequences will not be negative'.[6] Despite the 13 per cent scepticism across A/NZ, it is certain that the impact of climate change will have a profound effect on our lives and how we live them. In a climate change world we must accept that any action (adaptation or mitigation) will force human societies to change. How we interpret the changes needed will depend on how we understand, know, and live within our landscapes and environments. Our environment can teach us how to adjust to the coming changes by observing how it changes, why it changes, and what we can do to live with it. The capacity to adapt will prove to be the biggest challenge and will have most impact on indigenous communities, impoverished peoples, and groups whose ways and means for living are inextricably linked to the environment. In particular Pacific peoples who are part of the environmental ecosystem through belief, values, knowledge, and practices will experience substantial challenges to lifeways.

This is one of two companion books that investigate the role of indigenous knowledge (IK) in minimising the impact of climate change. Bryant-Tokalau's book *Indigenous Pacific Approaches to Climate Change: Pacific Island Countries* seeks to portray what has taken place in neighbouring PIC to date. Bryant-Tokalau traces a history of Pacific environmental management since the 1950s with the establishment of regional organisations, the impact of the Pacific's difficult nuclear history on these organisations, and the shift to other concerns such as biodiversity, waste, and climate. Global institutional developments within the UN and other multilateral organisations had an enormous influence on the way that PICs responded to their many environmental concerns, and indeed caused much stress in terms of dealing with institutional demands, but the PICs also had some influence on global practice, particularly through alliances such as Small Island States and the UNFCCC. Bryant-Tokalau acknowledges that formal religion, spirituality, and the fundamental belief systems of the many, complex Pacific societies are intrinsic to the many ways that communities, individuals, and governments respond to the challenges of climate change. She states that 'many of [the tradition-based responses] evolved over centuries, and often are more appropriate than the current "technical fix" response to inundation, droughts and major storms'. Bryant-Tokalau examines the theme of global responses versus traditional practices (and the stressors placed on those). Some of her examples include the artificial islands approach with particular emphasis on the long-established Pacific practice of creating land (one key example is Kokoifou artificial island in Langalanga lagoon, Solomon Islands). This was done both as a response to shortage of living space, and for cultural, relational, and environmental reasons. Other examples are the migration and resettlement options for increasingly uninhabitable lands, and responses to flooding and increased hurricane activity. Bryant-Tokalau discusses the long history of such developments and traditional adaptations that communities have always made and still make to the changing environments. Her examples span case studies from Solomon Island, Vanuatu, and areas of Micronesia and Fiji. In all respects it appears that A/NZ has a lot to learn from the Pacific. As with the Pacific solutions and practices, Māori tribal groups here in A/NZ stand well placed to be key players in adaptation strategies, policies, and practices that are referenced through Māori/ Iwi traditional knowledge. The book then acknowledges that IK frameworks will form the foundation for understanding and adapting to the many climate change challenges that lay ahead for A/NZ.

Neither this book nor Bryant-Tokalau's book will be dealing with the scientific focus around climate change. Rather we accept the evidence for it, and instead will focus on the key areas that impact most particularly on the social and cultural factors of human society. The book will be supporting earlier work by Barnett and Campbell that challenged the climate change science-and-policy orthodoxy and moved the thinking to a wider social dimension.[7]

Because of the focus on the areas in the IPCC report, both books have chosen not to specifically discuss energy projects designed to reduce the reliance on fossil fuels. We do acknowledge however the importance of these strategies for the wider Pacific region. There is a vast amount of literature and research around alternative energy and fossil fuel reduction, whereas positive stories about adaptation measures that focus on Pacific challenges from sea-level rise and flooding are less prolific. Taking the lead from the IPCC, the case studies we have chosen to include in the two books will focus on existing (and potential) mitigation and adaptation measures that minimise the impact of sea-level rise and flooding.

The term 'Pacific Island Countries' refer to those that are on the whole governed by the indigenous peoples. Hence the examples come from PIC such as Fiji, Solomon Islands, and Tonga (among others). The responses and reactions to the impact of climate change are diverse and develop from the individual country's cultural, political, economic, and environmental landscape.[8] However they are also united in a common Pacific stance to keep climate change very high up on global political agenda. As a bicultural nation, in A/NZ IK (or Mātauraka Māori or MM) operates alongside non-MM knowledge frameworks. On one hand, MM forms the basis for Māori tribal social, cultural, economic, and environmental management of their landscapes, environments, and the ecological systems contained within (including human).[9] On the other hand, MM is incorporated into some government policy and action in ways that differ to the diverse experiences in the PIC.

I acknowledge that there is a wealth of research and action to limit the challenge from climate change happening amongst indigenous peoples across the world. In particular, the Arctic Climate Assessment Report (ACCR) provides information for all the countries that count the Arctic Circle as part of their territories.[10] The third chapter of the ACCR report acknowledges the way that the many groups have adapted to environmental change and the many 'challenges posed by geography and climate', over millennia. They have done this by using their knowledge of the environment based on their intergenerational connectedness with the landscapes. The observations of how the environment has changed also now apply to the added challenges from climate change.[11] The value of that knowledge in meeting the new challenge lies 'primarily within the group and culture in which it developed' and continues to develop with each new challenge. IK is experiential knowledge that develops intergenerationally and is tied to specific environmental regions and circumstances. Its accuracy is dependent 'on the uses to which the knowledge is put, not on external valuation'.[12] The ACCR chronicles Arctic peoples' recent engagement with climate change through observation of weather, seasons, wind, and sea fog, and behavioural changes in animals and insects.[13] Indigenous place names are considered an important aspect of geographical regions in the Arctic Circle countries. They describe many of the climate and weather-related events and are often multi-dimensional referring to the geography, 'movement of animals, community history, and historic and mythological events'.[14] Place names as environmental indicators help to document changes over time. Changes in land-use may be influenced by changes to the environment and the given name may no longer reflect the purpose for which it was originally given.[15] Nevertheless, the reason for the naming is evident within stories that describe experiences over time with the place.[16]

Likewise, in the Teso region of Eastern Uganda, the Iteso people have acquired detailed knowledge about the functioning of their immediate environment, including observations and insights into a wide range of issues, challenges, and changes.[17] In the Torres Strait too 'local knowledge is a valuable asset in observing and managing environmental change'.[18] The Northwest Indian Applied Research Institute published a report (2006) that details climate change and Pacific Rim Indigenous Nations. The authors argued that on a region-by-region basis, indigenous peoples hold a 'collectively powerful body of climatological knowledge holistically interwoven…as environmental knowledge of a culture'.[19] What this body of research indicates is the rising importance of listening to indigenous peoples and learning from them with regard to understanding the social and cultural impacts from climate change.

In particular there is a growing body of research that investigates the role of IK in strategies that combine mitigation and adaptation policies and actions. Nyong et al. describe the experiences in the African Sahel where the integration of both is 'not a completely new idea', and exists within long-held Sahel indigenous sustainable resource and land management practices.[20] While acknowledging this wealth of experience and knowledge that exits across the indigenous world, both this book and Bryant-Tokalau's book are unashamedly about the island Pacific nearest to, and including A/NZ. Therefore the IK frameworks referred to are those of the PIC, A/NZ included, that form part of the Polynesian Triangle. This book is set in A/NZ and the Māori Environmental Knowledge (MEK) frameworks being referred to begin and grow from the experiences and practices of the Māori tribal groups in our particular land and sea environs. The examples given are from A/NZ because the book is about the A/NZ response (or not) to the impact of climate change. The author's tribal affiliation is to the southern Māori tribal group, Kāi Tahu. Therefore many of the examples are South Island focused. This is not intended to infer that North Island tribes are complacent about climate change and North Island tribal examples will be cited where appropriate.

IK (as used in this book) provides a comprehensive record and understanding of the mechanisms and interlocking systems that make up the world in which we live. The knowledge and systems that come from this understanding (our world views) provide rules and instructions that people will utilise to benefit in socially, economically, environmentally, and culturally appropriate ways. An understanding of how we live in our world determines the way that humans adapt to challenges such as climate change.

Traditional ecological knowledge or TEK (as used in this book) refers to a knowledge framework or sets of frameworks that are location-based and steer the relationship between people and a specific place. Berkes explains that TEK 'is limited to more explicitly land-related knowledge and is considered a subset of the broader category of indigenous knowledge'.[21] It is dynamic and builds on experiences and relationships with a particular land and/or seascape. A key idea in a Pacific world view is that land and sea are regarded as a seamless extension of each other with complimentary ecologies. This idea influences the decisions made and practices carried out in order to do battle with the impact of sea-level rise and associated flooding. TEK is the term we will use in this book as it refers to a more specific set of IK that is referenced and informed by the relationships between people and a specific place.

The climate change debate is constantly couched within the terms 'mitigation' and 'adaptation'. I will discuss this in the context of A/NZ, and the measures taken so far, because this is how A/NZ also debates the issues and it is unavoidable. However, I also propose to approach the debate from an IK perspective and discuss ways to negotiate the impact of climate change through the changing states in our relationships with the ancestral elements. In an IK context there is no division between mitigation and adaptation. The IK-informed actions, management, and spiritual practices work together to ensure that relationships between all parts of an ecosystem and wider environment are interrelated and are in a constant state of transition (or renegotiation). The origin stories explaining each relationship are recalled through genealogical connections and include the ancestors who control the elements such as Tāwhirimatea, the ancestor in charge of winds, rain, and other climate-related elements. We cannot control these ancestors, but can negotiate ways to make sure we can live within the environment they create. For example, in the context of climate change the relationship with the climate elements has been damaged. We have entered a transitional period, which requires adjustments to the new circumstances we now face. We can harness the power of the wind to be our power; we can take a leaf out of Māui's book and harness the power of the sun. Both these adjustments lessen our reliance on fossil fuels and slow the release of greenhouse gas (GHG) emissions into the atmosphere. These and other adjustments we make help transition the relationship to a new state in which we use the elements to help rather than hinder.

The A/NZ book is divided into three parts. In Chap. 2 the terms 'mitigation' and 'adaptation' are defined as they apply to the debate around climate change in A/NZ and the Pacific. The wider Pacific is directly affected by the result of rising GHG and therefore to be a good neighbour, A/NZ has responsibilities and obligations to develop both effective mitigation and adaptation strategies. These cannot be developed independently of each other and need to work hand-in-hand to be long lasting and future focused. A discussion on the ancestral links between the PIC and A/NZ demonstrates how MEK has its origins in the wider Pacific through language, culture and whakapapa (genealogy). The closeness of A/NZ to its Pacific neighbours is an ancient and complex mix of all these things.

Chapter 3, *Traditional Environmental Knowledge and Climate Change*, considers the challenge from the IPCC Fourth report, which called for IK to be utilised in the climate change adaptation. This chapter contains a discussion on what is meant by IK and the more localised TEK. Both terms are at times contentious because of the connections with tradition-based practices, knowledge, beliefs, and values. It is the term 'traditional' that causes scepticism towards the validity of IK to sit alongside the Western science counterparts. This view 'presupposes that tradition is locked into an inflexible, prescribed set of values, knowledge frameworks and principles'[22] that do not allow for a dynamic and ever-changing world. The term tradition however, merely means that knowledge is informed through past experiential systems and processes that provide guidelines for working through new challenges and situations. It is knowledge built up over time and has the ability to change and adapt when new challenges present themselves. As such it provides for an intergenerational set of rules and practices that accumulate through the experiences of each generation.

The IPCC (2000) identified land-use changes as a key factor in increasing the amount of GHG into the atmosphere.[23] A/NZ land-use changes can be traced early from Polynesian settlement through to contemporary times. Over time, Māori prospered through adapting to what was for them a harsh and variable climate and topography to ensure cultural, economic, social, and environmental sustainability. The people shaped the cultural landscapes just as the people's cultural practices were shaped by the landscapes they found themselves in. From 1840 European settlement in A/NZ accelerated the modification of the landscape including the rivers, wetlands, coastal areas, and flood plains that have left long-lasting consequences and challenges for contemporary peoples. The history of these modifications (Chap. 4) emphasises economic growth and development at the expense of environmental and cultural growth. The growing awareness of the uneven development priorities has resulted in two key pieces of legislation that concern New Zealand's land and sea territories. These are the Resource Management Act, 1991 (RMA) and the Exclusive Economic Zone and Continental Shelf (Environmental Effects) Act, 2013 (EEZ). It is understood that these are only two of many policies, and legislation that impact on the shape and management of New Zealand's land and sea territories and resources. But, they are two key pieces for future climate change mitigation and adaptation for government agencies, local bodies, and Māori and as such they will be included in that context. Also discussed is the role of the Environmental Protection Agency (EPA) because this is the government agency in charge of overseeing the RMA, and the EEZ. The RMA, EEZ, and the role of the EPA will be analysed to see how well they recognise, include, and practice Māori Environmental Knowledge (MEK) within the legislative frameworks and policies that will be guiding future land and sea-use changes and opportunities; and how these will drive adaptation practices. The EPA is also in charge of oversight for New Zealand's key climate change response to date, the NZETS.

The second part of the book examines A/NZ part in mitigation and climate change adaptation with a growing call for mitigation and adaptation to work hand-in-hand.[24] Indigenous peoples' resource and land management practices have developed and implemented extensive strategies to reduce vulnerability from climate change in ways that 'add value to the development of sustainable climate change mitigation and adaptation strategies'.[25] These are rich in local content,'[26] and are planned with local people who best understand the environment they live and work in.

In A/NZ the emphasis has been on developing mitigation strategies and action to reduce GHG through developing an Emissions Trading Scheme. Chapter 5, *New Zealand and the Emissions Trading Scheme*, outlines New Zealand's contribution to reducing GHG emissions. New Zealand is a signatory to the Kyoto Protocol and is committed to reducing the GHG emissions. The commitment also requires New Zealand to assist developing countries to combat the challenge of climate change. As David Bullock noted, 'New Zealand's record on climate change has been poor.'[27] He states that while emissions in the European Community have fallen by 2.2 per cent over the period 1990–2006, New Zealand's have increased by 25.7 per cent. Just five other Kyoto Protocol Annex 1 countries are performing more poorly than New Zealand.[28] Dubbed 'emissions impossible'[29] for good reason the NZETS faces criticism for not including her biggest GHG emitter, the agriculture sector. Globally the agriculture sector is responsible for 40 per cent of all GHG emissions and in A/NZ this is 49 per cent of the total emissions. In order to meet her responsibilities to her relationship with her Pacific neighbours New Zealand needs to shift its status from a poor GHG reducer to one that takes climate change seriously. Chapter 5 includes a discussion around Māori tribal involvement with the NZETS and what the Kyoto Protocol means for Māori.

Chapter 6, *Aotearoa/New Zealand Adaptation Strategies and Practices*, investigates who has responsibility for what in the context of adaptation policy, strategy, and practice. The examples included here demonstrate how local initiatives are working to minimise the impact of sea-level rise and flooding on the coastal environment and resources. These include the recognition of adjustments to change, the current adaption capacity and future requirements, and adaptive management strategies being considered and in some cases implemented. In A/NZ any climate change is inseparable from issues linked to natural hazards management and sustainable development. Natural hazard management is the responsibility of local and central governments and Iwi (Māori tribal groups). Iwi territories take into account the coastal, marine, and land environments and all the subsequent interactions and interrelated relationships among them. The environment and ecosystems are inextricably linked into the whakapapa (genealogies) of Māori communities and the kaitiaki role (conservation and sustainable practices) that Māori have within their respective takiwā (territories). Thus any Māori adaptation measures and processes link with those of other communities and agencies in A/NZ. The established Iwi relationships and treaty-based partnerships with government and local councils (LC) become increasingly important then when thinking about climate change adaptation in A/NZ. Māori adaptation responsibilities involve creating balanced processes and solutions across the economic, social, cultural, and environmental context. These are invariably informed through Mātauraka-a-Iwi (Iwi knowledge frameworks) and couched in the values and beliefs that underpin all Iwi interactions within their respective takiwā. Iwi knowledge can be reinforced with that of non-Iwi knowledge frameworks and practices to ensure the most beneficial outcomes for all those who interact within Māori territories. The concept of co-production of knowledge also plays an important part in climate change action.

The final part of the book will draw the preceding chapters together and also develop a 'Where-to-from-here' discussion including the adoption of MM principles surrounding relationships (origins, management, and future sustainability) as a way forward. Chapter 7 will also review the thinking around the developing science for more innovative approaches to crop development, housing, and recognising blue carbon and its potential for inclusion into mitigation strategies (such as Emission Trading schemes) across the Pacific. The purpose of the blue carbon review is to emphasise that like other environmental challenges in the Pacific, challenges, changes, and impact of climate change are ongoing. This supports how Pacific peoples understand 'in profound and elegant ways that "we are all related"'[30] and that the land-sea interface is an ever-changing and challenging space in our lives.

A/NZ as an industrialised nation has responsibilities and obligations to be a good neighbour to other Pacific dwellers. In the context of A/NZ becoming a better neighbour to the PIC, this book seeks to discuss how A/NZ can benefit from the wider Pacific TEK practices and strategies already in place; and how A/NZ can utilise MEK to achieve substantial inroads into transitioning into the new environmental states we are now encountering (Fig. 1.1).

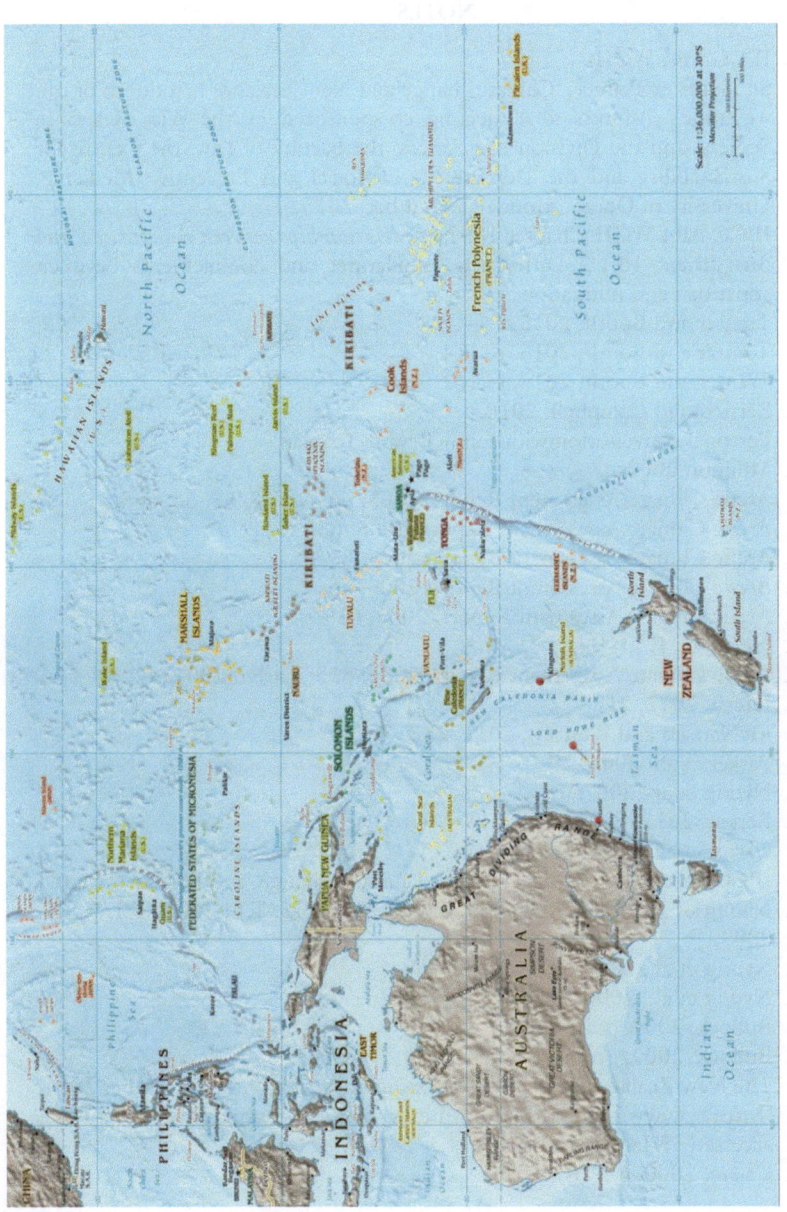

Fig. 1.1 Map of Oceania, *Central Intelligence Agency*, 2010. The 'Polynesian Triangle' comprises Aotearoa/New Zealand (in the south), across to Rapanui (Easter Island, in the east), and the third apex in the north, the Hawaiian Islands

NOTES

1. IPCC AR4 WGII.
2. Sir Geoffrey Palmer, 'Climate change and New Zealand: Is it doom or can we hope?' Address to a meeting co-sponsored by the Wise Response Society, and the Division of Sciences, the Faculty of Law, the Centre for Sustianability and the Department of Social and Preventive Medicine, University of Otago, Monday 5 October, 2015.
3. IPCC AR4 WGII Chapter 18: *Inter-relationships between Adaptation and Mitigation*. 18.1.2 Differences, similarities and complements between adpatioan and mitigation.
4. Tranter and Booth, 2015.
5. Tranter and Booth, 2015, p. 154.
6. Tranter and Booth, 2015, p. 162.
7. Barnett and Campbell, 2010.
8. Personal conversation with Jenny Bryant-Tokalau.
9. Mihinui, 2002, 1.
10. Arctic Climate Assessment Report, 2005, pp. 62, 63, 65, 66, 655.
11. Arctic Climate Assessement Report, 2005, 63.
12. Arctic Climate Assessment Report, 2005, 65.
13. Arctic Climate Assessemnt Report, 2005, 66.
14. Arctic Climate Assesment Report, 2005, 655.
15. Helander, E., 1999.
16. Arctic Climate Assessment Report, 2005, 655; Carter, 2004a.
17. Egeru, 2012, 217.
18. McNamara and Westoby, 2011, 887.
19. Parker et al., 2006, 29.
20. Nyong et al., 2007, 787.
21. Berkes, 2012, p. 9.
22. Carter et al., 2011, p. 19.
23. IPCC Special Report on Landuse, Land use change, and Forestry, 2000.
24. Nyong et al., 2007; Klein et al., 2003; Wilbanks, T.J., 2005; Tol, R.S.J., 2005.
25. Nyong et al., 2007, 787.
26. Nyong et al., 2007, 787.
27. Bullock, 2009, p. 2.
28. Bullock, 2009, p. 2.
29. The New Zealand Herald – *NZ's Emissions Impossible*, 16 December 2014. 'http://www.nzherald.co.nz/news/print.cfm?pbjectid=11374647', accessed 15 January 2015.
30. Cajete, 2000, p. 178.

INTRODUCTION 13

REFERENCES

Arctic Council. (2005). *Arctic Climate Assessment Report*. New York: Cambridge University Press.
Barnett, J., & Campbell, J. (2010). *Climate Change and Small Island States: Power, Knowledge and the South Pacific*. London: Earthscan.
Berkes, F. (2012). *Sacred Ecology* (3rd ed.). New York: Routledge.
Bullock, D. (2009). *The New Zealand Emissions Trading Scheme: A Step in the Right Direction?* (Institute of Policy Studies Working Paper 09/04, March 2009). Wellington: School of Government Studies, University of Victoria.
Cajete, G. (2000). *Native Science Natural Laws of Interdependence*. Santa Fe: Clear Light Publishers.
Carter, L. (2004a). Naming to Own. Place Names as Indicators of Human Interaction with the Environment. In *AlterNative. An International Journal of Indigenous Scholarship*, issue *1*, 7–25. Auckland: Nga Pae o Te Maramatanga/ The National Institute of Research Excellence in Māori Development, University of Auckland.
Carter, L. (2004b). *Whakapapa and the State. Some Case Studies in the Impact of Central Government on Traditionally Organised Māori Groups* (Unpublished PhD Thesis). Auckland: University of Auckland.
Carter, L., Kamou, R., & Barrett, M. (2011). *Literature Review and Programme Report. Te Pae Tawhiti Maori Economic Development Porgramme*. Published Report for Nga Pae o Te Maramatanga, University of Auckland.
Egeru, A. (2012). Role of Indigenous Knowledge in Climate Change Adaptation: A Case Study of the Teso Sub-region, Eastern Uganda. *Indian Journal of Traditional Knowledge, 11*(2), 217–224.
Helander, E. (1999). Sami Subsistence Activities – Spatial Aspects and Structuration. *Acta Borealia, 16*(2), 7–25. https://doi.org/10.1080/0800389908580495.
Klein, R. J. T., Schipper, E. L., & Dessai, S. (2003). Integrating Mitigation and Adaptation into Climate Change Development Policy. In N. Stehr & H. von Storch (Eds.), *Environmental Science and Policy, 8*(6), 579–588, December 2005.
McNamara, E. K., & Westoby, R. (2011). Local Knowledge and Climate Change Adaptation on Erub Island, Tores Strait. *Local Environment, 16*(9), 887.
Mihinui, B. (2002). Hutia to rito o te harakeke. A Flaxroot Understanding of Resource Management. In M. Kawharu (Ed.), *Whenua. Managing Our Resources* (pp. 21–33). Auckland: Reed Books Ltd.
Nyong, A., Adesina, F., & Osman Elasha, B. (2007). The Value of Indigenous Knowledge in Climate Change Mitigation and Adaptation Strategies in the African Sahel. *Mitigation and Adaptation Strategies for Global Change, 12*, 787–797. https://doi.org/10.1007/s11027-007-9099-0.
Parker, A., Grossman, Z., Whitesell, E., Stephenson, B., Williams, T., Hardison, P., Ballew, L., Burnham, B., & Klosterman, R. (Eds.). (2006). *Climate Change*

and Pacific Rim Indigenous Nations. Washington, DC: Northwest Indian Applied Research Institute (NIARI), The Evergreen State College, Olympia.

Tol, R. S. J. (2005). Adaptation and Mitigation: Trade-Offs in Substance and Methods. In N. Stehr & H. von Starch (Eds.), *Environmental Science and Policy, 8*(6), 572–578.

Tranter, B., & Booth, K. (2015, July). Scepticism in a Changing Climate: A Cross-National Study. *Global Environmental Change, 33*, 154–164. http://www.sciencedirect.com/science/article/pii/S095937801500758. Accessed 15 July 2015.

Wilbanks, T. J. (2005). Issues in Developing a Capacity for Integrated Analysis of Mitigation and Adaptation. In N. Stehr & H. von Storch (Eds.), *Environmental Science and Policy, 8*(6), 541–547.

CHAPTER 2

Setting the Scene

Abstract Mitigation and adaptation are defined in this chapter in the context of how they fit within the debate and action here in Aotearoa/New Zealand. Several indigenous scholars argue that these should not be thought of separately, but as one set of challenges. This is a perspective that best suits indigenous knowledge (IK) beliefs and practices as a way of dealing with the impact from climate change. As a Pacific Island nation and one apex of the Polynesian triangle, Aotearoa/New Zealand has its own IK frameworks, which have origins in the Pacific. The Pacific-Māori links will be discussed, as will the origins of Māori ecological knowledge within Aotearoa/New Zealand.

Keywords Māori and Pacific networks • Mitigation • Adaptation • Sea-level rise • Flooding

AOTEAROA/NEW ZEALAND AND CLIMATE CHANGE

The human impact on the climate is creating long-term climatic changes that will ultimately change the way we live over an extended period of time. The Earth's temperature has already risen by 0.4 per cent since the 1970s. Currently it is heading for 1.5 per cent warmer than it should be under naturally occurring conditions. In the 2015 Paris climate change agreement,[1] A/NZ committed to reduce GHG to 30 per cent below the

© The Author(s) 2019   15
L. Carter, *Indigenous Pacific Approaches to Climate Change*,
Palgrave Studies in Disaster Anthropology,
https://doi.org/10.1007/978-3-319-96439-3_2

2005 levels. Much has been said about the relevance of the A/NZ effort if the agriculture sector is not included in the emissions trading scheme. In 2016 New Zealand's Parliamentary Commissioner for the Environment, Jan Wright, expressed concern over the agriculture sector omission. She was not confident that planned technological developments would address the agricultural emissions and advocated for increasing carbon sinks—primarily through tree planting; 'it might not be the whole solution, but a million hectares of trees would make a big difference'.[2]

The debate around climate change contains a number of terms and an understanding of what is actually meant by climate. The weather is the term for events that occur on a short-term basis (daily, weekly) and over time build up a pattern of climatic behaviour for specific regions (e.g. rain, winds, sunshine hours). The observable patterns build to develop a long-term picture of the climate. Changes to the climate happen over time, including the naturally occurring release of carbon into the atmosphere. It is now known that certain human activities have produced the changes to naturally occurring GHE, by releasing unnaturally occurring carbon emissions into the atmosphere. Many of these changes such as widespread deforestation, destruction of wetlands, burning of fossil fuels, and changing land-use patterns (such as the forest to dairy conversion here in A/NZ) are contributing to the accelerated GHE. Scientists believe that changes on these scales could produce a number of results such as sea-level rise, melting sea ice and retreating glaciers, increased extremes in weather events, and increased surface temperatures.[3]

Some of these results from climate change can already be observed globally and within A/NZ such as retreating glaciers, flooding, increased drought conditions, and more frequency of violent storms. In the wider Pacific region, the changes such as sea-level rise, out-of-season hurricanes, and flooding are also observable changes. For the PIC many of these changes are largely not of their own making, but are as a result of larger industrialised countries' GHG emission rates over decades. The battle to reduce GHG emissions is mitigation; one of the two terms synonymous with climate change discussions. Mitigation strategies can be thought of as 'procedures or activities that help prevent or minimise the process of climate change',[4] decrease emissions of GHG, and increase the number of carbon sinks to naturally absorb GHG,[5] such as reforestation and improve/protect wetland environments.

The second most common term in the climate change debate is adaptation. Along with this is an understanding of adaptive capacity and adaptive management. The 2014 IPCC report defines adaptation as a process of adjustment to actual or expected climate and its effects. The need for adjustment necessitates assessing the appropriate adaptive capacity for vulnerable resources and landscapes, as well as 'the ability of a system to use these resources effectively in pursuit of adaption'.[6] This is managed through developing technologies, infrastructure, and flexible decision-making that will accommodate future challenges requiring adjustments to adaptive processes, practices, and systems. Kwadijk et al. refer to this as recognising adaptation tipping points. Should an environment move beyond the predicted tipping point, the current management strategy and /or processes may no longer work and an alternative will need to be developed.[7] The anthropogenic changes to the composition of the earth's atmosphere have altered the ecological balance among vulnerable environments. Recognising the future impacts that sea-level rise and flooding will have on these environments is part of the adjustment process that will need to occur in determining successful adaptation management systems, planning, and processes. The flexibility for future adjustment and capacity is inherent within an IK framework because fundamental to IK is managing relationships.

In A/NZ a key principle within MEK is that humans are part of any environmental system. As such they are an integral part of the ecology for the many resources needed to sustain the cultural and social norms of Iwi society. Through MEK, everything is personified so there is a supreme ancestor in charge of all elements of the world, including human. The supreme ancestor is the origin of the various participants within the natural world, and as such appears within the relationship structure that is expressed through the concept of whakapapa (genealogies). If a relationship is abused or ignored to benefit one above the other, then there are consequences. For example, if one partner in a relationship moves to gain control over the other, the dominant partner will change the relationship to an unequal state (tipping point). The subservient partner can either give in and subject to the new state or work to rebalance the relationship to a more mutually beneficial one. In IK and MEK all challenges to status quo are considered through how we initiated, managed, and maintained relationships. By considering their place in the world through a series of relationships (whakapapa) with ancestors and their descendants, Māori tribal groups can negotiate a set of rules for engagement (tikanga) and the appropriate cultural practices to enact them. Should an imbalance occur in the order of things, there are methods available to alleviate the damage (e.g. a rāhui, or prohibition on gathering specific resources) as the relationship between the resources, habitat, and people is renegotiated through a transitional period from one state to another. There is flexibility built into the decision-making process and within the tikanga system to adjust (adapt) the rules for engagement to meet the subsequent challenges and move forward.[8] The renegotiation of practices to survive the transitional period allows people to adjust to living within a changed environment. This practice of tiakitaka (guardianship, care, and responsibility) holds the key to ensuring that social, cultural, environmental, and economic factors are intergenerational.

It is acknowledged that in the present world there are other outside influences that have joined with the established relationships that Iwi members have with the environmental systems within their lands and waterways. For example the challenges that accompanied European colonisation included the loss of land and access to resources. These have had long-reaching impacts on the ability for Iwi to successfully manage these. It has required huge adjustment and adaptive management practices in order to rebalance the relationships Iwi have with their lands, rivers, estuaries, and coastal territories. In some cases the challenges have been insurmountable. The cultural, social, and environmental systems have reached a tipping point that left to breach has meant loss of important mahika kai sites and the knowledge that went with them. (This will be discussed further in Chap. 6.) Many of these challenges echo the experiences of our Pacific Island neighbours with whom we are forever linked through past whakapapa relationships.

New Zealand Māori and Pacific Networks and Connections

From the times of early voyages across our Oceanic highways, PIC have been adapting to many forms of environmental change. Adaptation strategies and practices are profoundly influenced by culture and cultural landscapes that have endured across time. These strategies and practices demonstrate means and ways of dealing with the contemporary and future challenges from climate change. Cultural landscapes are linked directly to how we know, understand, and work within them, so consequently any changes will also directly impact on the culture and the community. It is important that the changes and adjustments will ensure the culture remains true to its beliefs, practices, and values. These in turn underpin processes utilised to meet the challenges from climate change. In the Pacific, the adjustments to changing environmental states have been informed over time from historic environmental challenges and disasters. Many historic solutions come from generations of understanding and living with the environment and the understanding that humans are part of the ecosystem—genealogically, physically, and spiritually.

New Zealand Māori share many of the same Pacific values and cultural identity, which are inextricably linked to the land, seas, rivers, and mountains that make up the landscape. These shared beliefs and practices began though a wider ancient network of relationships that spread across the Pacific. The ancient links between Māori and Pacific peoples can be found in the many Māori tribal sayings, such as the following example:

> I found a great land covered with high mists in Tiritiri-o-te-moana, the open sea that lies to the south. (Kupe's discovery)[9]

This saying was attributed to the great Pacific voyager and navigator, Kupe, who is credited with sighting the great land mass to the south and naming it Aotearoa—land of the long white cloud. Kupe's sailing instructions were passed down to succeeding generations, which enabled peoples to voyage here and settle on the land. The Māori whakatauki (saying) refers to Māori people as '*te kakano i ruia mai i Rangiātea/ the seed which was sown from Rangiātea*',[10] which emphasises the early connections to the Pacific. Like their Pacific cousins, the Māori efforts will be underpinned by the values, practices, and beliefs that ensure cultural integrity is maintained. In the context of climate change, Māori will need to investigate ways to ensure that they can meet the challenge head-on.

MĀORI ECOLOGICAL KNOWLEDGE: LAND AND SEA

Māori Iwi (tribal groups) transmit MEK through traditional stories around the origins and deed of ancestors. King et al. describe the stories as representing the 'totality of experiences of generations of Māori in Aotearoa ... MEK also expresses the context within which it was developed and realised – that is, by Māori in Aotearoa'.[11] Another important concept that lies at the base of many Iwi philosophies and MEK is the notion of tākata whenua. The word tākata means people and whenua means land/placenta: combined they mean people from the land. Ngā Puhi tribal elder Māori Marsden described it as man being 'an integral part therefore of the natural order and recipients of her [mother earth] bounty'.[12] Land is a parent to people and as with the social obligations between parent and child, there are social obligations towards the earth and resources.[13] This is referred to as kaitiakitaka (guardianship, care, and responsibility), and the exercising of kaitiakitaka underpins the relationship between people and land. The atua who are associated with the various other parts of the environment (flora and fauna) as well as the oceans and waterways are also recorded within MEK and the whakapapa (genealogies) of the people. As King et al. remind us, all these associations and subsequent relationships link Māori people to the land and sea right here in A/NZ.

The term tākata whenua also refers to places of origin for each group's distinct whakapapa (layers of relationships; genealogies). The place of origin is always the central point for recalling associations and relationships. The specific place of origin for each tribal group is the tūrakawaewae: literally a standing place for the feet. The term contains both the practical (ira takata) and the spiritual (ira atua) elements that make up each living thing. In Māoritaka everything has a spiritual and physical component and these must always be in balance to ensure sustainability takes place—whether they be cultural, social, economic, or environmental contexts.[14] The people in residence at the tūrakawaewae are referred to as the ahi kā (the warm fires). The ahi kā ensure the connections are kept alive within the place itself. They maintain the hau or the breath of life of the place and everything that resides within it. They carry the role of the kaitiakitaka over land and environment. This requires understanding the integrated systems and processes that maintain the balance between humans and environment. From a Māori perspective this concerns all social, cultural, environmental, and economic contexts. Hence the practices and beliefs must work towards the most beneficial outcomes for any situation that requires resource use. The tools for accomplishing this are based in experiential knowledge acquired over generations.

For Māori the importance of land as the connecting influence to identity is a key part of the MM thinking. In a contemporary context Māori-owned land is governed and managed in particular ways under the Te Ture Whēnua Act, 1993.[15] There are three general categories of Māori Land: Māori freehold land which has constantly been in Māori hands; Māori customary land, which is land held in accordance with tikaka Māori; and General Land, the land owned by five or more (related) Māori people. The Act allows for five categories of land type, the most common being the Ahu Whenua Trusts. These are 'intended to promote and facilitate the use and administration of the land in the interests of the owners'.[16] In the context of climate change the way that Māori/Iwi manage their land holdings is related to the understanding of land as an integral part of their whakapapa; as according to tikaka (rules and practices for engagement); and with the intent of ensuring social, cultural, economic, and environmental priorities are met. The same applies to Māori/Iwi sea-based resources. MM is the process Māori/Iwi engage with the NZETS and how they understand the implications of Kyoto policy on Māori land-use.

The 2014 IPCC report emphasised that the two strategies (mitigation and adaptation) work hand-in-hand and should be implemented together to ensure that both short- and long-term impact is dealt with and effective.[17] In the past, A/NZ had yet to show commitment to adaptation strategies and practices in any substantial way. With the change of New Zealand Government in 2017, there has been more acknowledgment of the need for adaptation strategies and planning. A government report (expected in March 2018) will outline future scenarios and possible short- and long-term strategies. The present climate change minister, James Shaw has emphasised the urgency to minimise the impact from rising GHG through mitigation processes and at the same time, finding ways to adjust to the actual and expected changes. For this to happen Māori need to be a visible and vocal partner in the debate and subsequent activities because Māori are the knowledge holders of a system that will benefit how A/NZ can lessen the impact. MEK established Māori within the Pacific networks and relationships that can now be drawn upon for examples and strategies that have utilised IK. Māori, as the indigenous peoples in A/NZ, are in a better position to fully comprehend and learn from the lessons coming from the Pacific. This potentially places them in a leadership position in lessening the impact from climate change in A/NZ and moving forward within a changed environment.

NOTES

1. The 21st meeting of parties to the UN Framework Convention on Climate Change met in Paris in November/December, 2015 to (among other discussions) set new targets for reducing GHG emmissions.
2. Jan Wright, Parliametnary Commissioner for the Environment, quoted in the *Otago Daily Times* report, 'Work urged now on agricultural emissions', Thursday October 20, 2016, p. 14.

3. Definition of Climate Change in *EcoLife dictionary: A guide to Green Living*, p. 1, http://www.ecolife.com/define/climate-change.html, last accessed 4/09/2015.
4. Nyong et al., 2007, 791.
5. IPCC Fourth Assessment Report, 2004.
6. Kwadijk et al., 2010.
7. Kwadijk et al., 2010.
8. Carter and Ruru, 2005.
9. Peter Buck, *Vikings of the Sunrise*, 267.
10. Mead and Grove, 2001, 30.
11. King, Skipper and Tawhai, *Maori environmental knowledge of local weather*, 387.
12. M. Marsden and M. Henare, *Kaitiakitanga: A definitive introduction to the holisitc worldview of the Māori*, Wellington: The Ministry for the Environment, 1992, 16–17.
13. Letter to George Grey from Metapere Te Wai-puna-a-hau, Ngāti Awa, Grey New Zealand Manuscripts 453, Auckland University Archives, Auckland University Library.
14. Carter and Ruru, 2005.
15. Harmsworth, 2003, 10–11. The Te Ture Wēnua Act replaced the Māori Affairs Act, in 1993, with the primary purpose to keep Māori land in Māori ownershipand control.
16. Harmsworth, 2003, 11.
17. IPCC AR4 WGII Chapter 18: *Inter-relationships between Adaptation and Mitigation*. 18.1.2 Differences, similarities and complementaries between adaptaion and mitigation. https://www.ipcc.ch/publications_and_data/ar4wg2/en/ch18-1-2/, last accessed 2/6/2016.

References

Carter, L., & Ruru, J. (2005). "Freeing the Natives" The Role of the Treaty of Waitangi in the Reassertion of tikanga Māori. In N. Thomas (Ed.), *Te Tai Haruru. Journal of Maori Legal Writing*, 2, 15–36.

Harmsworth, G. (2003). *Maori Perspectives on Kyoto Policy: Interim Results. Reducing Greenhouse Gas Emissions from the Terrestrial Biosphere (C09X0212)*. Discussion Paper for Policy Agencies (Climate Change Office; MfE, MAF, TPK). Palmerston North: Landcare Research NZ.

Kwadijk, J. C. J., Haasnoot, M., Mulder, J. P. M., Hoogvliet, M. M. C., Jeuken, A. B. M., van der Krogt, R. A. A., van Oostrom, N. G. C., Schelfhout, H. A., van Velzen, E. H., van Waveren, H., & de Wit, M. J. M. (2010). Using Adaptation Tipping Points to Prepare for Climate Change and Sea Level Rise: A Case Study in the Netherlands. In *WiREs Climate Change*. John Wiley & Sons Ltd. http://wiley.com/climatechange

Mead, H. M., & Grove, N. (2001). *Ngā Pēpeha a ngā Tīpuna. The Sayings of the Ancestors.* Wellington: Victoria University Press.

Nyong, A., Adesina, F., & Osman Elasha, B. (2007). The Value of Indigenous Knowledge in Climate Change Mitigation and Adaptation Strategies in the African Sahel. *Mitigation and Adaptation Strategies for Global Change, 12,* 787–797. https://doi.org/10.1007/s11027-007-9099-0.

Traditional Ecological Knowledge in Climate Change

Abstract Parker et al. describe indigenous peoples as being resilient in meeting past challenges to their world views and lifeways. Utilising 'traditional strengths' (Parker et al. *Climate Change and Pacific Rim Indigenous Nations*. Washington, DC: Northwest Indian Applied Research Institute (NIARI), The Evergreen State College, Olympia, 2016, 12) makes them well placed to meet the new challenge from climate change head-on. The IPCC called for including indigenous knowledge (IK) to underpin adaptation. This chapter discusses IK, traditional ecological knowledge, and the A/NZ indigenous framework, Mātauraka Māori. The origin stories of how the land and sea became interlinked with people and the way people have worked with the environment for mutually beneficial outcomes will be analysed. The way IK frameworks work through a system of relationships will be expanded here to further support the notion that mitigation and adaptation cannot be thought of as separate events.

Keywords Experiential knowledge • Traditional practices • Traditional beliefs and values • Role of ancestors, states of transition

IK is a holistic knowledge framework that transmits knowledge, institutions, values, and practices from one generation to the next. It is often orally transmitted and this has prompted calls for indigenous peoples to record their knowledge to ensure that the IK can advance 'without fragmenting the cosmic connectedness between land, culture and knowledge'.[1,2] The term indigenous knowledge is often used interchangeably with the term traditional ecological knowledge. Berkes differentiates the two by claiming that TEK is a subset of IK and is 'a broader category within which traditional ecological knowledge fits'.[3] IK is a wider appreciation of a cultural set of values and practices that can be usefully adhered to across a number of (generic) situations.

TEK is transmitted through language that verifies a group's connection and claim to a place. This is done through dialect, visual imagery, deed of past ancestors, and the stories explaining the symbolic nature of the landscape features within the people's territory or environment. Therefore TEK tends to be environmentally referenced and region-specific in terms of weather, soil types, plant growth, coastal environment, and fauna and flora. The activities of the people themselves, utilising their own local TEK, is based on observation and management of the environment over a long period of time and is 'an unheralded source of adaptive capacity'.[4] In the Pacific islands TEK and practices are grounded in the ancient knowledge that people are one with the land and the sea and it is this perspective that has been integrated into Pacific Islands' climate change adaptation across their respective regions.

In A/NZ Māori knowledge that comes from this long-held association with the A/NZ is called Mātauraka Māori. It contains a set of values and practices that can be attributed to the Māori way of understanding the world different to that of the later Pākehā arrivals. Mātauraka Māori is IK. There is however, more than one Māori tribe (Iwi) in A/NZ and each tribal group has its own way of understanding its particular territory within the wider land mass. The subset of knowledge (to take Berkes definition) is called Mātauraka-a-Iwi—knowledge of the tribe—and it relates specifically to the relationships between individual tribal groupings and the specific place they belong to. Wiremu Doherty refers to Mātauraka-a-iwi as grounded in an 'environmentally-located base' that creates inseparable strands between people and the environment they live in.[5] The integration of people and environment develops mutual understanding and collective practices that grow through the generations, thus giving us Māori ecological knowledge (MEK)—Iwi-specific knowledge for managing ecological relationships.

Fikret Berkes explains that traditional knowledge should be seen as a process rather than as content. He argues that 'Scholars have wasted ... too much time and effort on science vs. traditional knowledge debate; we should reframe it instead as a science *and* traditional knowledge dialogue and partnership'.[6] Berkes suggests that the generation of new knowledge be combined with what is already known both to science and to TEK.[7] In particular, the climate change debate offers opportunities for this combination to occur.

King et al. remind us that despite Western scientific progress in the areas of climate modelling and forecasting, 'there is increasing recognition that Western scientific ideas are not the only [knowledge systems]... [and] indigenous environmental knowledge systems are increasingly being recognised as alternative domains of knowledge'.[8] In practice this is already occurring within parts of the wider Pacific. In Vanuatu there is an initiative led by the National Meteorological Service that combines the traditional knowledge surrounding weather signs and prediction techniques with scientific information.[9,10] In A/NZ South Island, Kāi Tahu (in collaboration with NIWA) have produced resources outlining the traditional regional variations in weather forecasting utilising generations-old observations and practices.[11] In the Fourth Assessment Report (2007) the IPCC identified that local knowledge '[is] an important, yet, to date overlooked component of its previous work and assessments', and confirmed that 'it will focus on local knowledge regarding the impacts of climate change and variability' thus stressing the overlooked but highly significant value of local knowledge, particularly in the Pacific.[12] The IPCC Report also stressed that adaptation is context-specific and therefore there is no single approach for reducing risks that would be appropriate across all settings. That being the case, a location-specific approach to adaptation is more likely to ensure long-term benefits and be intergenerationally successful. This places emphasis on traditional knowledge, and local knowledge systems and practices as major resources that have yet to be consistently utilised within adaptation strategies and planning.

The IPCC reference to TEK as an important resource for adaptation to climate change highlighted the need for further research and discussion. Two workshops were planned to 'identify, compile and analyse relevant indigenous and local observations, knowledge and practices related to understanding climate change impacts, adaptation and mitigation'.[13] These were held in Mexico City (July 2011)[14] and in Cairns, Australia (March 2012).[15] They brought together indigenous and marginalised peoples, and various agencies such as the United Nations University, IPCC, Secretariat of the Convention on Biological Diversity (SCBD), United Nations Development Programme (UNDP), and the United Nations Educational, Scientific and Cultural Organisation (UNESCO).

The workshops aimed to develop ways to understanding vulnerability, adaptation, and mitigation using IK that was compiled from local and regional data. They sought to engage with indigenous peoples, marginalised populations, and developing country scientists and encourage sharing of their research or knowledge. A key aim was to provide policy makers with the knowledge and data on the 'vulnerabilities, knowledge, and adaptive capacity'[16] of indigenous peoples and marginalised populations. The expected outcomes were more about providing information based on the shared knowledge and data gathered through the workshops. The main objective was to provide a 'broad international network'[17] of contacts with access to the collated database of TEK-based adaptation and mitigation strategies and/or case studies. Regardless of the motivations and the outcomes for the workshops, the real value comes from the recognition that TEK frameworks and practices will play an important and pivotal role in adaptation and mitigation for climate change. This is particularly relevant for the small island nations, including in the Pacific where TEK plays a prominent role. An important point also raised by the Indigenous People, Marginalised People and Climate Change (IPMPCC) workshops is that IK no longer exists on the margins of other knowledge frameworks. Rather it has become part of parallel knowledge pathways moving with those of other nations and cultures.[18,19] A merger of parallel knowledge frameworks can be described as co-production of knowledge where a number of knowledge systems interact to find solutions to common (mutual) challenges. When promoting the use of TEK in adaptation efforts, the IPCC noted that 'increasing such forms of knowledge with existing practices increases the effectiveness of adaptation'.[20] The co-reliance on both TEK and practices along with scientific knowledge and practices will produce integrated knowledge systems that provide future solutions to contemporary problems. This will become more important as resources are contested both within, and from without, the Pacific regions.

TEK AND THE PACIFIC

In the PIC, the traditions stress the belief that land and sea are a seamless extension of each other, with land having risen from the sea floor into the world of light. The traditions teach about the collaborative relationships and obligations between people and the land; people and the sea; land and sea. Indigenous island dwellers consider they are both land and sea-based peoples where the 'land and sea flow into each other, particularly in the in-shore, coastal zones where seascapes and landscape merge'.[21] The belief that 'land and sea are not discrete entities [and that] the one exists and only makes sense in relation to the other' underpins an IK perspective.[22] The traditions (or the 'rich cultural dimension'[23]) take the form of stories, songs, and traditional sayings that reinforce the generations of observation, practice, and listening to the environment. They 'encode and impart…ecological and cultural knowledge'[24] about the world and everything's place within it. It is experiential knowledge and increases intergenerationally over time—hence the location-specific characteristic of IK. The state and conditions of the environment along with the symbols of weather, land/sea conditions, and the life cycle of the entities that live within them will determine periods of plenty, periods of scarcity, and the periods of preparation towards these. The traditions (regardless of genre) are interlaced with the exploits of ancestors and atua (ancestors, supreme beings) and it is through these exploits that the knowledge is learnt and the practices and rules are set. The genealogical layers of relationships (whakapapa) acknowledge the way that generations of people contribute to the knowledge and practices that are foretold in the traditions. Weather patterns, stars, cloud formations, sea conditions (e.g. wave patterns, water appearance, conditions of the tides), and fauna (e.g. flowering of plants and trees, eating and breeding patterns of the birds) work together to herald particular weather patterns and signs for seasonal food and activities that ensure the relationship between human and environment is a mutually beneficial arrangement. Māori whakatauki (Māori sayings) emphasise the relationships between different parts of the environment, such as references to particular stars as signs for plentiful harvests and weather conditions. There are some key ancestral figures whose exploits, descendants, and roles are inextricably linked to the environmental knowledge that surrounds the land and the sea across the Pacific. Two of these are Takaroa and Tane. Takaroa is well known across the Pacific as the ancestor associated with the sea and all living things within it. In Kāi Tahu tradition he is also the first husband of Papatūanuku (the earth). In some Pacific traditions, such as Tonga, Tagaloa has other responsibilities. He is known to dwell in the heavens as the god of thunder and lightning, and god of carpenters, arts and inventions. He is also responsible for fishing up the Tongan islands (Fig. 3.1).

In the land ecosystems (forests and birds) the supreme ancestor in A/NZ is Tāne—the god credited with separating the parents Rakinui (the sky father) and Papatūanuku (the earth mother) thus allowing light into the world. Māori whakatauki express the dominion of Tāne as '*Te wao tapu nui a Tāne* (The great sacred forest of Tāne').[25]

Fig. 3.1 Photo of
Takaroa. Carved pou
(post) at Warrington
Beach, Dunedin.
(Carved by members of
Te Whare Wananga o Te
Whānau Arohanui,
Watiati, Dunedin. The
spirals represent the past,
present, and future
generations of kaitiaki
for the resources around
the East Otago
coastline—the domain of
Takaroa. It stands at the
interface between sea
and land)

The rules for engagement with the forest dictate that Tāne's permission must be sought before a tree is cut or any living thing is taken from the forest.

Takaroa and Tāne have an uneasy alliance. The relationship is constantly shifting and re-adjusting most markedly where land meets sea—the estuaries, coastline, lakes, and river mouths. The two atua are thus jointly responsible for the areas that are most susceptible to sea-level rise, flooding, and storm surges and it is the stories that surround these two that will inform the knowledge of how best to adapt to any challenges that arise. The pairing of Tāne and Takaroa is apparent across the Pacific. In Hawaii the pair are known as Kane and Kanaloa and are described 'in legend as cultivators, awa drinkers, and water finders, who migrated from Kahiki and travelled about the islands'.[26] Kanaloa is associated with the underworld and is mentioned in a chant in which Hawaii 'is fished up from the very depths of Kanaloa'.[27] Kane is described as the 'bringer of food plants to Hawaii' and as god 'of plant and animal life'.[28] In all regards the realms of Takaroa and Tāne; Kanaloa and Kane; and the other Pacific variants of the atua are inseparable. This reinforces beliefs that land and sea are seamless extensions of each other reliant on the goodwill of either atua for survival of the accompanying ecosystems. Further evidence to support this comes from the stories that explain the origin of land through phenomena that Patrick Nunn referred to as land being either fished up or thrown down.[29]

The Origins of Land: Fishing Up

Accompanying the stories about Takaroa and Tāne, further stories inform the origin of land. In most, a fisherman pulls the sea floor above the water line to become the many Pacific islands. Some stories have Tagaloa (in Tongan tradition) as the fisherman. In many cases the land-fisher is Māui who is a common thread across the Polynesian Pacific. In Hawaii Māui's exploits vary from region to region, but with the one common theme explaining why the islands are separate and not one whole land mass. In the version from Hawaii's Eastern region Māui catches a large fish, but a mistake from his brothers causes the fish to sink back into the depths of the sea. This is the reason that the Hawaiian Islands are not joined as one land.[30]

In the Hawaiian Kauai version Māui is tasked with drawing the islands together to make one land mass. Māui catches a large fish, but ignores his mother's warning not to take a certain bailer into the canoe. The bailer turns into a beautiful woman and as the crowds cheer the sight, Māui's brothers are distracted and the fish escapes. The islands then pull apart and remain so to this day.[31] In Niue, Māui pulls a large white punga (rock) to the surface and it becomes the island of Niue.[32] Traditions from other Pacific Islands also tell the story of Māui as fisherman. The North Island of New Zealand for example is known as Te Ika a Māui (the fish of Māui). In the story Māui's brothers do not allow him to go fishing with them and so he hides away in the canoe. Once at sea Māui reveals himself to his brothers and produces his fishing hook made from the bones of his ancestor. He smears the hook with his own blood and proceeds to hook and land a great fish, 'Behold. The fish of Maui is raised up, a land-fish, the earth…their canoe lay on dry land.'[33] The land as fish and Māui as the land-fisher can also be found in chants from Aitutaki[34] and Rakahanga,[35] and stories from Mangareva and Tokelau.

In Samoa the name for the land-raiser is Ti'iti'i, which recalls one of the names given to Māui across other parts of the Pacific and refers to his origins—*Maui a Tiketike o Taranga* (Māui of the topknot of Taranga). In Vanuatu the land-raiser is named Wuhngin.

Across the Pacific, Māui continues to fish. In 2014 a new island appeared in the Tongan island group, and the 2016 Kaikoura earthquake resulted in a sea floor rise between 0.5 and 2 m along parts of A/NZ Kaikoura coast.

ORIGINS OF LAND: THROWN DOWN

Nunn also discusses the notion of lands being thrown down due to one of the many atua dropping soils or rocks into the ocean.[36] Nunn lists several examples such as this one from Kadavu Island (southern Fiji), where soil was spilled from a basket of earth stolen from Nabukelevu volcano as the thief was running away. A similar story comes from the Marshall Islands when earth is spilt from a basket as the atua, Etao, is flying through the air; and '[in] another account, all the islands in the Marshall and Caroline Island groups of Micronesia were carefully placed in the ocean from a basket of earth'.[37] While not as numerous as land-fisher stories, in at least one area (Aotearoa/New Zealand), islands are formed from both methods: North Island is fished up and South Island is thrown down as the following story illustrates.

> Takaroa's grandson, Aoraki and his brothers ventured out in Aoraki's great canoe to visit Takaroa's wife, Papatuanuku. Following the visit, they were to return to their own realm, but a mistake was made in reciting the lifting karakia (incantation) and the great canoe crashed down, resting on an underwater reef. Aoraki and his brothers scrambled to the high side of the canoe that was still free from the water and waited there. Over time they turned to stone and Aoraki who was sitting the highest became the highest point in New Zealand's South Island (Aoraki/Mt Cook) In time, his grandson Tūterakiwhānoa came looking for him and discovered what had happened. After mourning the loss of Aoraki and his brothers, Tūterakiwhānoa set about fashioning the great canoe into the land now known as Te Wai Pounamu/South Island.[38]

The story has synergy with the fishing up stories, as it is from the canoe of Aoraki that Māui fished up his great fish, Te Ika a Māui/North Island.

The idea of land being either fished up from the ocean floor, or thrown down, further establishes the Pacific-wide knowledge that land and sea are one—there is no separation between the land above water and that below. In some stories there are either people living on the fish as it emerges from the water, or people were created from the soils of the land. Hence people and land were created and emerged together into the world of light (Te Ao Marama).[39]

Understanding the realms of Tāne and Takaroa and the complexities of their relationship is helpful in understanding the ecosystems that form the interface between sea and land. The coastal areas of tidal estuaries and marshes, mangroves, seagrass meadows, and river mouths are important areas for not only offering natural barriers to storm surges and sea-level rise, but as being among the most 'intense carbon sinks on the planet'.[40] These areas are referred to as blue carbon areas because of the carbon being stored in coastal vegetation ecosystems (biomass and sediments). This will be examined in more detail in Chaps. 6 and 7, through discussion around traditional beliefs between land and sea within the modern context of carbon sinks.

TEK: Climate and the Weather

Another key set of traditions revolves around the weather and the forces that control it. In the traditions from A/NZ, Tāwhirimātea is in charge of the winds and weather patterns. The knowledge of Tāwhirimātea's activities has built up over time, thus giving us access to climate knowledge for specific regions. He works often against Takaroa and Tāne, but also assists one against the other. Storms ravage coastlines and inland areas when the wrath of Tāwhirimātea is awakened, thus decimating the domains of both Takaroa and Tāne. Storm surges combine with Takaroa and leash destruction upon the coastlines, and winds and rain work against the land to create flooding and damage to vegetation. MEK tell us that Tāwhirimātea is expressing his displeasure at Tāne for being responsible for the separation of Papatūanuku (the Earth) and Ranginui (the Sky). Tangaroa too has some influence here as the first lover of Papatūanuku and has vested interest in her well-being or otherwise. In Niue the god involved in good weather patterns is Leomatagi who captured the winds and placed them in a cave, and it is Tapakaumatagi who rules the winds. In Tahiti, when Ra'a (sometimes Ra-patia) is angered he sends destructive westerly winds from the island of Borabora; and in Lau, Fiji it is Vinaka who sends winds that cause mortal death.[41]

Regardless of the name of each controller of winds and tempests, the relationships built between the atua are mitigated through various processes that help to reduce the impact from the atua rivalries. This helps to retain a balance in the natural ecosystems and provides lessons and experiences in overcoming environmental challenges from less-than favourable short-term weather conditions and long-term climate changes. The forces occasionally unite to remind us of the consequences of messing with nature. These often manifest in violent weather patterns and are retold through TEK stories. What these stories are telling us is that the weather is forever uncontrollable and although may be predictable from observation and experiences, it will always be a dominant force in our lives. The weather cannot be changed, because it is a daily, short-term variable pattern. Climate on the other hand is long-term and there is strong evidence of how human-induced influences can change the climate over the long term and mess with the environment. Some changes are too great to completely overturn. Yet in a long-term situation such as climate change an accommodating yet uneasy alliance can be negotiated between the atua and the peoples who inhabit their respective realms.

Integrating Western Science into TEK

Berkes claims that 'local knowledge can supplement the explanatory power of global climate change models, and provide grounded information on the actual impacts'.[42] In order to do this Davidson-Hunt et al. note that 'Coproducing knowledge in response to environmental change requires new institutional arrangements that provide community control, meaningful collaboration and partnership, and significant benefit sharing.'[43] Co-production of knowledge will ensure that adaptation will occur through complex interactions among knowledge systems. Most importantly, this would be informed, from a PIC perspective, by the country's TEK and practices. Over time and generations, the practices are reworked when changing circumstances come into play. This provides a history of solutions and practices to be referred to when new challenges occur. Davidson-Hunt et al. refer to changing circumstances as a process of adaptation that has been forced upon 'aboriginal societies' in the context of colonisation and global natural resource markets.[44] In a context of resistance to outside acculturation practices and policies, 'aboriginal peoples' have adapted to change by 'incorporating knowledge, practices and technologies consistent with … customs and values'.[45] This has been achieved through developing relationships and networks that, in line with Pacific perspectives, may be useful immediately or at a later time. One example is the A/NZ Māori concept of whakawhanaukataka (managing relationships). This concept provides guidelines for forming and sustaining relationships over time. This generally relies upon a complex system of reciprocal exchanges that form the basis for the ongoing relationship.

The concept, whakawhanaukataka, is also useful for understanding the relationship and network structures formed between indigenous and non-indigenous communities, organisations, and businesses. The knowledge acquired from external relationships will allow new technologies, values, and institutions to be incorporated into the TEK frameworks. At this point they become part of the community's response to environmental change. It would be naïve to say that indigenous peoples have prior or traditional knowledge about climate change as it is being defined today. What they have instead is 'sensitivity to critical signs and signals from the environment that something unusual is happening'.[46] The responses to earlier changes have been incorporated into the culturally defined adaptation measures over time. These observed changes and the measures put in place to cope with them are examples of the evolving TEK across the Pacific. It is clear then that 'place-based research and local observations have a crucial role to play in environmental changes'[47] including climate change.

It becomes apparent then that the integral relationships between people and their environment are at the heart of environmental adjustment and disaster adaptation. The key ancestral connections are with the atua who control the elements within land, sea, and climate and the way land-use has been managed within the environment they create. Changes to land-use have had major influences on the relationships and in contemporary A/NZ; legislative controls have influenced the way land-use is managed.

NOTES

1. Posey, p. 39.
2. Carter, 2003, 2004a; Sillitoe, 1998; Berkes, 2012; Posey, 2002.
3. Berkes, 2012, 9.
4. McNamara and Westoby, 2011, 887.
5. Doherty, 2010.
6. Berkes, 2012, xxiii.
7. Berkes, 2012, xxiii.
8. King, et al., 2007.
9. *Combining Traditional knowledge and Meteorological forecasts in the Pacific to increase community resilience to extreme climate events*, The Secretariat of the Pacific Regional Environment Programme (SPREP), www.sprep.org/pein/samoa.
10. Chand et al., 2014.

11. *Te Waipounamu weather and climate forecasting*, NIWA Poster Series No. 2017-3 (Te Reo Māori version) and 4 (English language version), funded through the National Science Deep South Challenge. For further information and to see the list of Kāi Tahu kaumātua (elders) who contributed knowledge to the project, see www.niwa.co.nz.
12. McNamara and Westoby, 2011, 888.
13. Report: Indigenous People, Marginalised Populations and Climate Change (IPMPCC), *Vulnerability, Adaptation and Traditional Knowledge 19–21 July 2011, Mexico City*, 'http://indingeous%20knowledge%20and%20climate...ples,%20marginalised%20Climate%20climate%20change.html', 1 accessed 17/2/2015.
14. International workshop on *Indigenous Peoples, Marginalised Populations and Climate Change: Vulnerability, Adaptation and Traditional Knowledge* – Mexico City, Mexico 19–21 July 2011, 1.
15. International workshop for *Indigenous Peoples, Marginalised Populations and Climate Change, Adaptation, Mitigation and Traditional Knowledge*, 26–28 March 2012, Cairns, Australia.
16. IPMPCC report, *Indigenous Peoples, Marginalised Populations and Climate Change*, 2011, 4.
17. IPMPCC report, *Indigenous Peoples, Marginalised Populations and Climate Change*, 2011, 4.
18. Carter, 2014.
19. Davidson-Hunt et al., 2013.
20. IPCC WGII AR5 Report *Climate Change 2014: Impacts, Adaptation, and Vulnerability. Summary for Policy makers*, 2014, p. 23.
21. Fuary, 2009, 32.
22. Fuary, 2009, 33.
23. Fuary, 2009, 33.
24. Fuary, 2009, 32.
25. Brougham et al., 2009, 59.
26. Beckwith, 1970, 63.
27. Beckwith, 1970, 61.
28. Beckwith, 1970, 61.
29. Nunn, 2003.
30. Beckwith, 1970, 231.
31. Beckwith, 1970, 232.
32. Cowan, 1923, 238–43.
33. Thornton, 1992.
34. Buck, 1954, 52.
35. Buck, 1954, 122.
36. Nunn, 2003, 360.
37. 2003, 360.

38. The story is a summary of various 'tellings' about the origins of Te Wai Pounamu, A/NZ South Island that are commonly known to Kai Tahu tribal members.
39. Carter, 2014.
40. Vierros, 2013.
41. Craig, 1989.
42. Berkes, 2012, 190.
43. Davidson-Hunt et al., 2013, 44.
44. Davidson-Hunt et al., 2013, 1.
45. Davidson-Hunt et al., 2013, 1.
46. Berkes, 2012, 172.
47. Berkes, 2012, 191.

REFERENCES

Beckwith, M. (1970). *Hawaiian Mythology.* Honolulu: University of Hawaii Press.
Berkes, F. (2012). *Sacred Ecology* (3rd ed.). New York: Routledge.
Brougham, A. E., Reed, A. W., & Kāretu, T. (2009). *The Raupō Book of Māori Proverbs.* Auckland: Penguin Group (NZ).
Buck, P., (Te Rangihiroa). (1954 [1975 reprint]). *Vikings of the Sunrise.* Christchurch: Whitcomb and Tombs Ltd.
Carter, L. (2003). The Bureaucratisation of Genealogy. In *Ethnologies Compares* (Vol. 6). Paris: University of Montpellier. http://alor.univ-montp3.fr/cerce/r6/1.w.htm. Last accessed 2 July 2003.
Carter, L. (2004a). Naming to Own. Place Names as Indicators of Human Interaction with the Environment. In *AlterNative. An International Journal of Indigenous Scholarship*, issue *1*, 7–25. Auckland: Nga Pae o Te Maramatanga/ The National Institute of Research Excellence in Māori Development, University of Auckland.
Carter, L. (2004b). *Whakapapa and the State. Some Case Studies in the Impact of Central Government on Traditionally Organised Māori Groups* (Unpublished PhD Thesis). Auckland: University of Auckland.
Carter, L. (2014, December 4–7). *We Are Not Drowning, Pacific Identity and Cultural Sustainability in the Era of Climate Change.* Conference presentation, Pan Pacific Indigenous Resource Management, Panel 3 – 'Managing the Sea and Land in Times of Climate Change', at Pacific History Association Conference, Taipei and Taitung, Taiwan.
Chand, S., Chambers, L., Waiwai, M., Malsale, P., & Thompson, E. (2014). Indigenous Knowledge for Environmental Prediction in the Pacific Island Countries. *American Meteorological Society.* https://doi.org/10.1175/WCAS-D-13-00053.1.

Cowan, J. (1923). The Story of Niue. Genesis of a South Sea Island. *The Journal of the Polynesian Society, 32*(128), 238–243.

Craig, R. D. (1989). *Dictionary of Polynesian Mythology*. New York: Greenwood Press.

Davidson-Hunt, J., Julian Idrobo, C., Pengelly, R. D., & Sylvester, O. (2013). Anishinaabe Adaptation to Environmental Change in Northwestern Ontario: A Case Study in Knowledge Coproduction for Non-timber Forest Products. *Ecology and Science, 18*(4), 44. http://www.ecologyandsociety.org/vol18/iss4/art44

Doherty, W. (2010). *Mātauranga a Tūhoe. The Centrality of Mātauranga-a-iwi to Māori Education*. Unpublished PhD Thesis. University of Auckland. http://hdl-handle.net/2292/5639.

Fuary, M. (2009). Reading and Riding the Waves: The Sea as Known Universe in the Torres Strait. *Historic Environment, 22*(1), 32–37.

King, D. N. T., Skipper, A., & Tawhai, W. B. (2007). Māori Environmental Knowledge of Local Weather and Climate Change in Aotearoa-New Zealand. *Climate Change*. Springer Science + Business Media BV, 385–409. https://doi.org/10.10077/s10584-007-9372-y.

McNamara, E. K., & Westoby, R. (2011). Local Knowledge and Climate Change Adaptation on Erub Island, Torres Strait. *Local Environment, 16*(9), 887.

Nunn, P. D. (2003). Fished Up or Thrown Down: The Geography of Pacific Island Origin Myths. *Annuls of the Association of American Geographers, 93*(2), 350–364.

Parker, A., Grossman, Z., Whitesell, E., Stephenson, B., Williams, T., Hardison, P., Ballew, L., Burnham, B., & Klosterman, R. (Eds.). (2016). *Climate Change and Pacific Rim Indigenous Nations*. Washington, DC: Northwest Indian Applied Research Institute (NIARI), The Evergreen State College, Olympia.

Posey, D. (2002). Upsetting the Sacred Balance. Can the Study of Indigenous Knowledge Reflect Cosmic Connections? In P. Sillitoe, A. Bicker, & J. Pottier (Eds.), *Participating in Development. Approaches to Indigenous Knowledge* (p. 39). London: Routledge.

Sillitoe, P. (1998). The Development of Indigenous Knowledge: A New Applied Anthroplogy. *Current Anthropology, 39*(2), 223–252.

Thornton, A. (1992). *Maui tiketike-a-taranga:* Te Rangikaheke to Grey. *GNZMMSS, 43*, 896–973; Auckland Public Library Collection. This version is also reprinted in Agathe Thornton, *The Story of Maui by Te Rangikaheke*. Edited with translation and commentary by Agathe Thornton. Christchurch: University of Canterbury Maori Studies.

Vierros, M. (2013, October 10). Communities and Blue Carbon: The Role of Traditional Management Systems in Providing Benefits for Carbon Storage, Biodiversity Conservation and Livelihoods. *Springer Science+Business Media Dordrecht, Special Edition: Climate Change*. https://doi.org/10.1007/s10584-013-0920-3. Last accessed 2015.

CHAPTER 4

Aotearoa/New Zealand and Land-Use Changes

Abstract Aotearoa/New Zealand has a long history of land-use changes modelled on remembered practices from the Pacific, and the challenging environmental conditions early Māori faced. Post 1840 the increase in Pākehā settlers saw an accelerated level of land-use changes such as wetlands drainage and degradation, deforestation, agriculture, and coastal developments. The consequences of the past changes have new meaning under climate change conditions, such as the increase in GHG emissions. In a contemporary context there are two key pieces of legislation that now control land-use change here in A/NZ: The Resource Management Act, 1991 (RMA) (and amendments) and the Exclusive Economic Zone and Continental Shelf (Environmental Effects) ACT, 2012 (EEZ). The discussion will focus on provisions in the Acts and their effectiveness from Māori perspectives.

Keywords RMA • EEZ • EPA • Māori and legislation • Land-use changes

When Polynesian explorers first discovered the lands in the southern Pacific they were met with unfamiliar terrain and climate variations which varied from temperate in the north of the North Island to sub-alpine in the south of the South Island. Many of the food staples that came with the early voyagers would not thrive here and some only if soils were artificially modified to improve survival rates. The early Polynesian arrivals set about reshaping the landscapes, which occurred at the same time as the Polynesian culture reshaped itself to become distinctly Māori. Many of the traditions and practices stemming from these times form the basis for the experiential knowledge and practices that are MM. Evidence of these can be found in the many place names that cover both land and sea. They trace the early landing and settlement places and internal migrations that resulted from overcrowding, resource depletion, and/or war for domination and territory. Many of the names mark resource-gathering sites (mahika kai) that are still in use today. The sea-based names mark, for example, fishing reefs, travelling routes, tidal currents, kelp beds, fishing grounds, and tribal boundaries. The notion of land and sea as one can be found in the place naming because landmarks (such as mountains) are triangulated with other land and sea-based landmarks to indicate the position of fishing grounds. As one Kāi Tahu taua (woman elder) remarked, 'we knew the land from the sea – before we even travelled inland.'[1] As the cultural nuances became apparent across landscapes and within the social organisation and practices, adaptation for economic, social, and environmental development also increased. These can be found in the archaeological record that plots the traditions across the landscape.

Archaeological gardens and settlement sites are the most common ways early settlement is seen within the changing landscapes. Evidence of climatic and environmental adaptation can be readily found. The variable climates found in the new islands meant the growing season for crops such as kumara (sweet potato) was much shorter here. They also encountered seasons—seasons of plenty but more noticeably seasons of scarcity. This required new and innovative cultivation methods to ensure successful yields and steady food supplies over the periods of scarcity. Janet Davidson noted how 'alteration of soils, by the addition of charcoal, sand, gravel, pebbles and shell fragments, seems to have been a feature of Maori horticulture practice', in order to improve drainage, fertility, and quality.[2] Davidson also states wetland drainage was used to enlarge garden areas and the development of excavated kumara storage pits in some parts of North Island.[3] The storage pits were a uniquely A/NZ innovation because a shorter growing season meant that crops grown from seed had no time to reach ultimate harvest conditions. The storage pits ensured that harvests could be kept free from damp and pests during the colder months, and also provide a stock of tubers for sprouting and planting for the following year's crop. In areas where kumara could not grow (beyond Bank's Peninsula in the South Island Canterbury region) other alternative starch sources needed to be found. The south Canterbury place name Te Umu Kaha (Temuka) describes the place where vast earth ovens were excavated to produce the kouru (sugars) from the Tii trees (New Zealand Cabbage trees). The Tii trees' fibrous shoots were also eaten. Māori utilised all the resources available (land and sea based) many of which required innovation and experimentation to be useful. The adaptation strategies and practices stemming from changing climatic and environ-mental conditions developed over time and each new adjustment was informed by experiences found in past knowledge and practices.

Māori, like their Pacific cousins, created artificial land and examples can be found in archaeological sites across the Hauraki Plains (Te Ika a Māui/North Island) where land was artificially raised above the swampy ground. Geoff Park describes the artificial land creation at Oruarangi, Hauraki Plains area, where the ground was built up to offset wet living conditions. In settlements along the Waihou river, can be found 'tens of thousands of cubic meters of sub-fossil and midden shell, sand, silt and clay, to raise [the land] above flood level'.[4] Park observed that he began to see Oruarangi as a '*built* place' that had required massive human effort to keep it 'from a river that frequently became a vast inland sea'.[5]

Other modifications included burning forests and land clearances for both encouraging fern root plantations and increasing cultivation areas. There are a number of examples of vast areas of land reshaped to form hilltop pā (fortified villages) sites with many terrace levels and storage pits for food and water.

Once European settlement began in earnest, the acceleration in land-use changes was evident. Wetlands were drained to increase productive farmland and the resulting loss of habitat has had long-lasting impacts. Some rivers were diverted to increase usable land for agriculture and horticulture, and utilised for farmland irrigation. Other rivers have also been dammed to provide hydroelectricity sources for the spread of indus-try and urbanisation. Streams and creeks around towns and cities were channelled through culverts and often disappeared underground during the urban land expansions.

Large areas of landscape were reshaped for urban settlements and for the materials contained within it. Gold, coal, and in the case of some of Auckland's volcanic cones, scoria and other metals were quarried for road and building projects. Opencast gold mines are still productive in the Central Otago area and in Waihi, Bay of Plenty. On the South Island western coast, coal mining is a main source of employment and economic growth (though this has slowed in recent years).

Forestry was, and remains, a large industry in the North and South Island for both native and exotic forests. Since the Forestry Amendment legislation in the 1990s, native forests can no longer be milled, but at least in one area there is provision for storm-damaged logs to be extracted. This occurs with the Māori-owned lands, which were legislated to Māori owners in the 1906 South Island Natives Lands Act. As in the past when native forests were cleared for agricultural use, exotic forests have recently been cleared for the same purpose. Large areas of the previously forested land between two central north island towns, Rotorua and Taupo, have been turned into large-scale dairy farms. Other regions, which were not traditionally used for dairy farming, are now being converted to dairying such as rural lands in Southland. The increase in dairying across New Zealand has had detrimental impacts on New Zealand's fresh water resources, and has created a corresponding increase in GHG.

Although this is a brief outline of New Zealand's historical land-use, it stands to demonstrate the ever-changing landscape here in A/NZ.[6] The modifications were generally for increased economic gain and Māori too benefited from historic and contemporary land-use changes to increase tribal holdings and economies. All of these have changed the landscape and created modified environments that have influenced the long-term modification of the climate. Draining of wetlands for example destroys natural carbon sinks. As wetlands are cleared or degraded, the carbon stored in them is released and the 'continued clearance of blue carbon ecosystems contributes significantly to atmospheric greenhouse gases'.[7] This contributes to the increased incidence of GHG here in New Zealand as more wetland areas are drained for agriculture, urban expansion, and industry. Forest clearances also reduce carbon sinks although reforestation is being recognised in the New Zealand ETS with carbon credit trade-offs made available for reforestation projects. Reshaping the contours of the land increases erosion and meddling with flood plains creates future flooding disasters. Coastal housing developments destroy fragile sand dune ecologies as does land reclamation over mangrove areas, such as has happened in the Auckland motorway system.

Land-use changes have altered and modified the natural carbon capture and storage (CCS) that is situated within the soils, wetlands, mangroves, coastal vegetation, and the 'offshore coastal wetlands [that] give way to expansive areas of seagrass, kelp beds and unvegetated seabed'.[8] As noted by Crooks et al., 'these ecosystems reflect a progressive transition from the land to the open continental shelf and the ocean beyond'.[9] In A/NZ development now takes into account the social costs from change as well as the environmental and economic. A/NZ is a bicultural nation so the cultural impact for Māori is also factored in. To control the speed and breadth of development, legislation and policies from both central and local governments were introduced. These are designed to create managed development that recognises that people are part of the environment too and that a balance between people's needs and those of the environment must be sought. While it is not the intention to reproduce all of the numerous polices, legislation, and strategies here, there are two notable pieces that are relevant to land-use changes, and changes to the way we utilise our coastal marine environment. One is the Resource Management Act, 1991 (RMA) and the other is the Exclusive Economic Zone and Continental Shelf (Environmental Effects) Act, 2012 (EEZ). The two Acts contain evidence of intent to incorporate MEK and Māori consultation policies within the legislation and the policies implemented by the Crown agency, the Environmental Protection Agency (EPA) that administers the legislation. The Acts will be summarised to see how they account for any future climate change adaptation policies or what consideration is given to climate change. The RMA and the EEZ outlined here are specifically designed to guide local councils and regional councils in developing strategies and long-term plans for future land-use changes and are by no means the only legislations and polices in place. The role and activities of the EPA will also be summarised in context of its administration of both the RMA and EEZ, and later in Chap. 5, the NZETS. The key piece of climate change mitigation legislation, The Climate Change Response Act, 2002, will be discussed in this chapter.

THE RESOURCE MANAGEMENT ACT, 1991

Perhaps the most well-known Act in New Zealand's plethora of legislation is the RMA. The Act is important because 'it provides the basis for local authorities to design rules for sustainable management of natural and physical resources'.[10] It means that no one (industry, companies, or individuals) can make any changes to do with land without first obtaining permission unless it is a 'permitted activity within a local authority plan'.[11] The RMA is the key piece of New Zealand legislation that initiates a process for notification and planning consents for land-use changes.

The RMA is particularly important in triggering Iwi consultation processes for any changes to lands and waterways within Iwi territories. There are two sections in the RMA that specifically address MEK. Section 6(e) deals with provision for matters of national importance and for Māori this is 'the relationship of Māori and their culture and traditions with their ancestral lands, water, sites, wāhi tapu, and other taonga'.[12] This insists applicants acknowledge MM and Māori world views in all physical and spiritual contexts. The RMA 1991 provides a framework for partnership between Local Government New Zealand (LGNZ) and the Iwi/Māori who reside within the local authorities' regions. The second part of the RMA 1991 that specifically concerns Māori is Section 7(a). This section deals with the MM concept of kaitiakitaka. The rules for carrying out the practice of kaitiakitaka vary regionally according to locally expressed MEK, but the overall intentions remain the same—the protection and sustainable management of resources including land and marine. Section 8 asks that decision-makers take the principles of the Treaty of Waitangi into account—partnership, protection, and participation. The principles recognise that firstly Iwi/Māori are the treaty partners; secondly that they enjoy full and meaningful consultation and participation. When the first two principles are fully realised then cultural integrity, values, practices, and knowledge are maintained to ensure full recognition and protection to MEK.

The legislation guides LGNZ in the Iwi consultation and participation process for matters pertaining to local resource management. The LGNZ organisation conducted two surveys (1997 and 2004) that investigated the levels and scope of engagement between local councils (LC) and Māori. The two reports showed that over time there was an 'increase in the levels of engagement' and improved implementation of formal and informal consultation processes.[13] Although the intent is there, it is doubtful if Māori/Iwi are experiencing a true bicultural management of A/NZ natural resources.[14] Other Māori commentators and Iwi have also expressed concerns about the ability of RMA to truly deliver bicultural partnership and management of natural resources.[15] The key issues that plagued the RMA were around the nature of consultation and an adequate joint decision-making process; and the capacity for both Iwi and LC to participate fully in environmental management.[16]

The 2003 Resource Management Amendment Act introduced two sections that were said to address the inadequacies of the original Act. Hutchings noted that the two sections that propose joint management options (s36B and 36E), have the potential to 'open space for Iwi/local government partnership', but lack of resourcing and LC willingness could ensure it remained a 'theoretical possibility'.[17] She comments that while the amendment has potential to create opportunities for Iwi and local authorities, there is also the risk that lack of capacity among local authority staff to fully understand MM may create 'inequitable interpretation and implementation', options that favour remaining with Western values and knowledge frameworks.[18] Māori have continued to remain sceptical that the RMA amended version would allow them to effectively exercise their rangatirataka and kaitiakitaka. The RMA sits within a raft of policies and legislations that control land-use within A/NZ that do not always favour MM views and practices. These tend to be underpinned by the way the two dominant political parties (Labour and National) think about, and practice their treaty partnership. The differing viewpoints creates what Hutchings has referred to as the 'ebb and flow of racial policies'[19] that ultimately determine the level and scope of MM within A/NZ resource management strategies, planning, and practice. The acceptance of MM as best practice will come from the innovation and creativity of Iwi/Māori and wider communities, who have the vision to see the value an MM knowledge framework brings to a bicultural management regime.[20]

One important tool that has come about since the 1991 Act is the Iwi Management Plans (IMP). Hutchings describes these as 'the most significant tribal planning tool' to provide Māori with a voice in resource management. The Plans allow LC to understand Iwi resource management agendas and best practice grounded in MM. Most importantly most Iwi have included the issues they face around climate change. The importance of the IMP was recognised in the RMA Amendment Act, 2003 where LCs are obliged to take the plans into account. This creates two immediate concerns: the capacity resourcing available to develop the plans, and the '(re) negotiation of the (bi) cultural basis of environmental management'.[21]

Sir Geoffrey Palmer recently analysed the RMA in the context of climate change. He noted that the 2003 amendment to the Act 'introduced provisions prohibiting consent authorities from considering the effects of greenhouse gas emissions on climate change when making rules to control discharges into the air; and when considering applications for a discharge permit'.[22] The consents and conditions to control the emissions were instead to be controlled through a National Standard, but as Palmer has noted, '[to date] no such standard has ever been promulgated...New Zealand's key environmental statute is disabled from considering what is a crucial issue relating to climate change'.[23] Consequently the primary environmental statute is 'deficient as a key mechanism for addressing climate change adaptation'.[24] Therefore the RMA does not provide the teeth needed for LCs to take action until it is amended to do so. Lack of action in this area has far-reaching and long-term consequences not only for New Zealand, but also for Pacific neighbours who are facing challenges that originated from industrialised nations like New Zealand.

While the RMA deals with land-use changes, the coastal and sea changes are monitored and controlled through a relatively new piece of legislation called the Exclusive Economic Zone and Continental Shelf(Environmental Effects) Act, 2012 (EEZ).

The Exclusive Economic Zone and Continental Shelf (Environmental Effects) Act, 2012

The land/sea areas subject to this piece of legislation are firstly the Exclusive Economic Zone (EEZ) defined as 'the area of sea, seabed and subsoil from 12 to 200 nautical miles offshore'; and the second area is the Continental Shelf (CS) known as 'seabed and subsoil of New Zealand's submerged landmass where it extends beyond the EEZ'.[25] The EEZ Act controls activity in the interface between land and sea and from where the RMA leaves off. In terms of the merger between land-based wetlands, rivers, and estuary areas, and the progression out towards the CS, this Act provides rules for use, control, and access. The types of activity managed through the Act include several activities that have an impact (directly or indirectly) on the acceleration of climate change factors in New Zealand's coastal and marine environments. These include gas, petroleum, and minerals extraction, CCS, and aquaculture.[26]

The EEZ is important in providing guidelines for the consent process in the above activities and includes acknowledgement of the potential impact on the CCS capabilities for our coastal wetland and marine ecosystems. The acknowledgement of CCS in this ecosystem environment is important as it should ensure that any further land-use changes that impact on the land-marine coastal interface will be monitored and controlled through this legislation. It is a key tool in the mitigation for climate change here in A/NZ and provides guidelines for planning and implementing adaptation measures and strategies that may result from sea-level rise and storm surges. The Environment Protection Authority implements the EEZ, and has developed extensive guidelines for consultation and engagement with Iwi/Māori that provides a Māori voice in any consent process.

Both the RMA and the EEZ legislation drive the management of the land/sea interface that mirrors the MEK notion 'ki uta, ki tai – from the mountains to the sea'.[27] Ki uta ki tai encapsulates both the ideas of mana whenua (power and authority) and kaitiakitanga over land, sea, and resources within these areas. While Iwi, hapū, and whānau communities and organisations manage and control the ki uta ki tai practices and processes, they need to be aware of the added restrictions and controls within the RMA and EEZ. These need to be mediated along with MEK approaches and activities that are recognised through the Acts. Within the EPA are a number of measures that help mediate MEK within non-Māori processes, organisations, and communities.

The Environmental Protection Authority

The EPA is a key government department in the control and management of changes to A/NZ's land and sea resources. The EPA's role includes national consenting under the RMA, management of the NZETS and the New Zealand Emission Unit Register, regulation of hazardous substances, new organisms, ozone-depleting chemicals, hazardous waste exports and imports, assessment of environmental effects in Antarctica, and managing the environmental effects of activities in the EEZ and CCS.[28] The management of hazardous wastes comes into play through local bodies that are charged with natural hazards within their regions and/or cities. The natural hazards area is also where climate change adaptation sits within the local body adaptation strategies and policies. Key partners in environmental management are other government agencies and departments, industry, science and research agencies, and Māori. The EPA has comprehensive guidelines for working with Māori, particularly the Māori consultation process. This is covered by the policy called *He Whetū Mārama*.

He Whetū Mārama is a framework that guides the EPA in understanding its statutory obligations under the Treaty of Waitangi to Māori. Therefore it includes reference to the three 'Ps'—protection, participation, and partnership. The three 'Ps' ensure that the EPA acts honourably, and in good faith in informed decision-making (partnership); takes positive steps to ensure Māori interests, knowledge, and experience are valued (protection); and effective engagement with and input from Māori (participation).[29]

The key components of this policy ensure that EPA fulfils obligations to fully informed decision-making that values MEK. This encourages full engagement with and input from Māori groups directly affected by any changes and opportunities surrounding A/NZ land and marine environments. On the surface then the EPA ensures that MEK is at the forefront of any engagement processes and outcomes. This is emphasised in a fourth Treaty of Waitangi principle that appears in the *He Whetū Mārama* policy—the principle of Potential. This principle recognises that EPA decision-making and activities impact on potential future Māori development and growth.[30]

The emphasis on MEK and engagement with Māori is also reiterated very strongly in *He Whetū Mārama* where EPA stresses that 'policy, processes and decision making is fully and effectively informed by Māori perspectives', and 'the EPA maintains relationships that ensure Māori are productively involved in its decision making and associated activities'.[31] The operational structure for EPA includes Ngā Kaihautū Tikanga Taiao—people who oversee and 'help EPA to incorporate Māori interests and concerns in its decision making'.[32]

It becomes evident then that these two key pieces of legislation and one government agency are there to ensure that MEK is included and utilised in decision-making for any changes, challenges and opportunities to the land and sea environs. This signals the intent to incorporate MEK in aspects of climate change mitigation and adaptation albeit in oblique ways. It also signals the intent of the EPA to make sure that international investors and organisations are fully aware of the requirements to consult fully with Māori and that indeed they are obliged to do so. The obligation to Māori and MEK within the RMA, EEZ, and the EPA policies and processes is quite clear. How well it operates though is not as clear. To support Crown-directed agencies in comprehensive engagement policies and practices, two South Island Iwi Rūnaka (multi-hapū district councils) have implemented policy documents of their own to ensure that, in particular, the oil and gas exploration and extraction industries are fully aware of the obligation to Māori.[33] Kāti Huirapa Rūnaka ki Puketeraki (KHR) and Te Rūnaka o Ōtākou (RoO) have engaged the Kāi Tahu ki Ōtākou[34] consultancy group to work with a committee of rūnaka members to produce the engagement document, *Kā Rūnaka expectations for the oil and gas industries*.[35] The document 'serves as a starting point' for oil and gas companies wishing to operate in Otago, especially in the rohe [region] of Kāti Huirapa Rūnaka ki Puketeraki and Te Rūnaka o Ōtākou. It is intended to assist oil and gas companies to engage in a meaningful way with Manawhenua in Otago, for the benefit of both parties.[36] The document is future focused and includes reference to a proposed climate change policy, particularly regarding increased degradation and disturbance of the blue carbon areas within the rūnaka takiwā. The Rūnaka engagement document insists that the principles of partnership, protection, participation, and potential are fully realised and the contexts in which these apply is very clear. The language used spells out clearly and plainly the expectations and the obligations for engagement and relationship building.

Of particular importance in the KHR takiwā (region) is the principle of ki uta ki tai that engages the inland puna (sources) for the Waikōuaiti river, the river's mergence into the Waikōuaiti estuary at Karitane, and the coastal fringes including kelp beds and shellfish beds; and the remnants of the coastal wetlands area, such as Hawkesbury Lagoon area. The wider catchment area surrounding the Waikōuaiti river is the subject of the Kāti Huirapa climate change project, 'Inaka[37] as an indicator of change. Future proofing inaka in the Waikouaiti river catchment area in the context of climate change'.[38] The project aims to develop a Mātauraka-a-Iwi model for ecosystem governance, management, and future sustainability in the context of climate change. The project sits under the umbrella of the New Zealand National Science Challenge: BioHeritage research, which is investigating ways to predict and manage resource and bioheritage tipping points across A/NZ. Although locally based, the model will be transferable to other regions and research projects within the Kāi Tahu South Island takiwā and beyond to North Island tribal groups. The Kāti Huirapa-led project is one of several region-specific climate change projects within the Te Rūnanga o Ngāi Tahu (TRoNT)[39] climate change strategy. The TRoNT climate change project (led by the TRoNT Strategy and Influence Team) contains both adaptation and mitigation approaches. Kāi Tahu realise that both need to be taken into account in order to develop effective, future-focussed plans and practices to reduce the impact from climate change within the South Island Kāi Tahu territories. The Strategy is also underpinned and driven by Kāi Tahu values and values-based practices that acknowledge the interrelated connections with landscape and ecosystems across Te Wai Poumanu. NIWA has produced a number of projected impact reports for the Kāi Tahu takiwā. These provide a summary of the impact within each of the 18 rūnaka regions so that Kāi Tahu will be informed about what is needed to support planning and future management 'from those who know their place or business the best'.[40] The first step for the TRoNT Strategy and Influence team is to develop a strategic direction based on 'the analysis of all the information received from NIWA, whānau, kaimahi and external sources to identify challenges within the Kāi Tahu takiwā and across the tribal network.'[41] In the South Island, Kāi Tahu are taking their role of kaitiaki seriously by taking control and leading the way to limit the impact from climate change across the tribal territories. The guiding principles couched in Kāi Tahu knowledge, values, and practices will underpin the future strategies and planning for short- and long-term development solutions.

In East Otago the KHR/RoO draft Oil and Gas engagement document supports the EPA insistence that Māori and MEK is fully included within any land/sea use changes. The KHR and RoO regional climate change strategies (as part of the overall TRoNT Climate Change strategy) will ensure future utilisation of MEK in South Island climate change policies and practice. Iwi-led initiatives such as the TRoNT Climate Change strategy provide a distinct Māori voice that takes the guesswork out of whom should be listened to and partnered with in environmental matters. It is a wait-and-see game as to how well people and organisations listen and take actions as prescribed in the government legislations and policies, and expected through the treaty partnerships.

The New Zealand Government's push for economic advantage from the minerals and wealth held within the seabed has overridden any effective and early engagement between industry and Māori. The lack of opportunity to fulfil one aspect of MEK—building long-standing relationships—has been tempered with the emphasis on economic gain. While the EPA insists upon appropriate engagement practices through its *He Whetu Marama* policy, there is still resistance to fully engaging with Iwi. This resistance comes from the belief that the social, cultural, and environmental factors will somehow impede economic gain. This leads us to the discussion around the New Zealand Government's key climate change mitigation strategy—the NZETS—that is also driven by a strongly economic agenda.

NOTES

1. Personal conversation with Professor Khyla Russell, Kāi Tahu, Kāti Mamoe, Waitaha Iwi (2003).
2. Davidson, 1992, 120–21.
3. Davidson, 1992, 120.
4. Phillips (2000) cited in McFadgen 2007, 166.

5. Park, 1995, 52–3. [Italics in original text].
6. For a comprehensive and authoritative source of land-use changes in A/ NZ see, E. Pawson and T. Brooking (eds), *Making a New Land. Environmental Histories of New Zealand* (Dunedin: Otago University Press, New Edition, 2013).
7. Vierros, 2013.
8. Crooks et al., 2011, 7.
9. Crooks et al., 2011, 7.
10. Ruru, 2013, 29.
11. Ruru, 2013, 29.
12. Ruru, 2013, 34.
13. Hutchings, 2006, 96; LGNZ, 2004.
14. Hutchings, 2006, 99.
15. Matunga, 2000; Ministry of the Environment (MfE), 2000.
16. Matunga, 2000; MfE, 2000; Hutchings, 2006, 98.
17. Hutchings, 2006, 99.
18. Hutchings, 2006, 100.
19. Hutchings, 2006, 103.
20. Hutchings, 2006, 104; Matunga, 2000.
21. Hutchings, 2006, 102.
22. Sir Geofrey Palmer, 'Climate change and New Zealand: Is it doom or can we hope?' *Address to a meeting co-sponsored by the Wise Response Society, and the Division of Sciences, the Faculty of Law, the Centre for Sustainability and the Department of Social and Preventive Medicine, University of Otago,* Monday 5 October, 2015.
23. Palmer, Monday 5 October, 2015.
24. Palmer, Monday 5 October, 2015.
25. Environmental Protection Authority, *Exclusive Economic Zone* (http:// www.epa.govt.nz/publications-resources/faqs), Accessed 25/09/2015.
26. Environmental Protection Authority, *Exclusive Economic Zone* (http:// www.epa.govt.nz/publications-resources/faqs).
27. Te Runanga o Ngai Tahu (http://www.ngaitahu.iwi.nz).
28. Environmental Protection Agency (http://www.epa.govt.nz/), last accessed 29 September 2015.
29. Environmental Protection Agency (http://www.epa.govt.nz/). Last accessed 29 September 2015.
30. Environmental Protection Agency (http://www.epa.govt.nz) Last accessed 29 September 2015.
31. Environmental Protection Agency (http://www.epa.govt.nz) Last accessed 29 September 2015.
32. Environmental Protection Agency (http://www.epa.govt.nz) Last accessed 29 September 2015.

33. Ruckstuhl et al., 2017.
34. Kāi Tahu ki Ōtākou (KTKO) is an environmental consultancy group employed by the three main Otago rūnaka: Kāti Huirapa Rūnaka ki Puketeraki, Te Rūnaka o Ōtākou, and Te Rūnaka o Moeraki. Since December 2017 it been known as Aukaha: a title that best represents the changing face of Iwi involvement in environmental management in Otago.
35. Ruckstuhl et al., 2017.
36. Ruckstuhl et al., 2017, 1.
37. Inaka (the young fish from a species of *Galaxias*) are commonly referred to as whitebait and are one of the mahika kai species in the Kāti Huirapa takiwā. The research project is part of the wider National Science Challenge research programme, BioHeritage, Theme 3; RA 3.1 – *Understanding tipping points from a Māori perspective*. The research team is Dr. Lyn Carter (Principle Investigator, Kāti Huirapa and University of Otago), Dr. Rose Clucas (Kāti Huirapa, independant researcher) and members of Kāti Huirapa Rūnaka.
38. The Inaka project is further discussed as a case study in Chap. 6.
39. Te Rūnanga o Ngāi Tahu (TRoNT) is the central governance and management body for the predominant South Island Iwi, Kāi Tahu. The TRoNT governance board is made up of the 18 regional rūnaka who oversee the tribal political, economic, social, and cultural development. Kāti Huirapa Rūnaka ki Puketeraki and Te Rūnaka o Ōtakou are the two rūnaka in the East Otago area including Dunedin City. For further information see 'www.ngaitahu.iwi.nz'.
40. Policy statement from the Te Rūnanga o Ngāi Tahu Strategy and Influence team who are charged with developing Kāi Tahu-wide strategies for dealing with the impact from climate change.
41. Te Rūnanga o Ngāi Tahu Strategy and Influence team policy statement to the 18 Kāi Tahu Rūnaka (tribal regional councils).

REFERENCES

Crooks, S., Herr, D., Tamelander, J., Laffoley, D., & Vandever, J. (2011). *Mitigating Climate Change Through Restoration and Management of Coastal Wetlands and Near-Shore Marine Ecosystems. Challenges and Opportunities* (Environment Department Paper 121). Washington, DC: World Bank.

Davidson, J. (1992). *The Prehistory of New Zealand*. Auckland: Longman Paul Limited.

Hutchings, J. (2006). Negotiating (Bi)cultural Environmental Management Under the Resource Management Act. In M. Mulholland (Ed.), *State of the Māori Nation: Twenty-First-Century Issues in Aotearoa* (pp. 95–105). Auckland: Raupo Publishers.

Local Government New Zealand. (2004). *Local Authority Engagement with Māori: Survey of Current Council Practices July 2004.* Wellington: Local Government New Zealand.

Matunga, H. (2000). Decolonising Planning: The Treaty of Waitangi, the Environment and a Dual Planning Tradition. In A. Memon & H. Perkins (Eds.), *Environmental Planning and Management in New Zealand.* Auckland: Dunmore Press.

McFadgen, B. (2007). *Hostile Shores. Catastrophic Events in Prehistoric New Zealand and Their Impact on Maori Coastal Communities.* Auckland: Auckland University Press.

Ministry for the Environment. (2000). *Reporting on the Key Issues and Good Practice Guidelines for Local Authorities to Better Understand Statutory Responsibilities.* Wellington: Ministry for the Environment.

Park, G. (1995). *Ngā Uruora (The Groves of Life). Ecology and History in a New Zealand Landscape.* Wellington: Victoria University Press.

Ruckstuhl, K., Gale, K., Carter, L., Ellisson, E., Flack, S., & Russell, K. (2017). *Kā Rūnaka Expectations for Oil and Gas Companies in East Otago.* Dunedin: Kai Tahi Ki Otago Ltd.

Ruru, J. (2013). Te Ture – Mineral Law and Māori. In K. Ruchstuhl, L. Carter, L. Easterbrook, A. R. Gorman, H. Rae, J. Ruru, D. Ruwhiu, J. Stephenson, A. Suszko, M. Thompson-Fawcett, & R. Turner (Eds.), *Māori and Mining.* University of Otago: Māori and Mining Research Team. http:/otago.ourarchive.ac.nz

Vierros, M. (2013, October 10). Communities and Blue Carbon: The Role of Traditional Management Systems in Providing Benefits for Carbon Storage, Biodiversity Conservation and Livelihoods. *Springer Science+Business Media Dordrecht, Special Edition: Climate Change.* https://doi.org/10.1007/s10584-013-0920-3. Last accessed 2015.

CHAPTER 5

Aotearoa/New Zealand and the Emissions Trading Scheme

Abstract Aotearoa/New Zealand's (A/NZ's) key piece of mitigation strategy is the New Zealand Emissions Trading Scheme (NZETS). The scheme was put in place to reduce greenhouse gas emissions in A/NZ in line with the Kyoto Protocols. This chapter will outline the development, implementation, and subsequent amendments to the NZETS, and the Māori reaction and compliance with it. Discussion will include plans for Māori land reforestation projects and the impact of the scheme from a social, economic, environmental, and cultural perspective. In question too is the tension between complying with an economic-focused mitigation measure, and MM practices, beliefs, and values that the NZETS ignores.

Keywords NZETS, Māori compliance • Kyoto Protocols • GHG emissions • MM vs. NZETS compliance

A/NZ is a member of the Pacific Islands Forum and thus has responsibilities and obligations to that relationship. In the context of climate change across the Pacific, A/NZ responsibility is to reduce the amount of GHG emissions that we produce. The PIC are facing a situation that is not wholly of their making and New Zealand is among the industrialised nations that have contributed to the situation through the emissions of GHG. New Zealand has also committed herself through the Kyoto Protocol to reduce GHG emissions and to assist developing countries to

© The Author(s) 2019
L. Carter, *Indigenous Pacific Approaches to Climate Change*,
Palgrave Studies in Disaster Anthropology,
https://doi.org/10.1007/978-3-319-96439-3_5

combat the challenge of climate change—her Pacific neighbours included. As David Bullock noted, 'New Zealand's record on climate change has been poor.'[1] He states that while emissions in the European Community have fallen by 2.2 per cent between 1990 and 2006, New Zealand's have increased by 25.7 per cent, with only five other Kyoto Protocol Annex 1 countries performing more poorly than New Zealand.[2] Dubbed 'emissions impossible'[3] for good reason the NZETS faces criticism for not including her biggest GHG emitter, the agriculture sector. Following the 2014 climate change meeting in Lima, Peru, the (then) Climate Change Minister, Tim Groser, said that New Zealand faced a big challenge in meeting the commitments to cut emissions.[4] To uphold her responsibilities to her Pacific neighbours, New Zealand needs to shift its status from a poor GHG reducer to one that takes climate change seriously to meet the Kyoto Protocol targets.

A separate discussion will focus upon the impact of the NZETS on Iwi and the relevance of the Kyoto Protocol to Iwi. The discussion is premised in the strong Māori economy, which was valued in 2010 at NZ\$42 billion with an annual growth rate of 4.5 per cent. It is a largely land-based economy with major interests in forestry, agriculture, tourism, and fisheries. From its very nature then, the Māori economy is part of the economic-focused NZETS programme; and Iwi/Māori are vocal within the climate change debates. Iwi share with PIC many of the same concerns, values, and cultural identity that are inextricably linked to the land, seas, rivers, and mountains. In the context of climate change, Māori too will need to investigate ways to formulate policy and strategy to ensure they can meet the challenge head-on.

A/NZ's two main islands make up a total land area of 27 million hectares. Of this land area 9 million hectares are in pastoral land and indigenous forests equal to 6.3 million hectares. The planted exotic forests make up 9 million hectares of the total land mass, with protected parks and reserves making up the remaining 5 million hectares. The A/NZ economy has been, and remains, agriculturally based. As such A/NZ is unavoidably exposed to the key risk areas of climate change identified in the recently released fifth IPCC report: sea-level rise, flooding, and wildfires. The IPCC report acknowledges that A/NZ has yet to put any significant adaptation measures into place while on the other hand A/NZ insists that it is committed to doing its fair share in reducing global GHG emissions.

A/NZ began its participation and contribution to reducing the production of GHG in 1988 when the Labour Government introduced the Climate Change Programme. This consisted of policy responses brought together under the Comprehensive Strategy on Climate Change (CSCC). The Strategy aimed to address sources and sinks of all GHG emissions and 'to maintain them at that level beyond the turn of the century'.[5] This called for a domestic target that returned net anthropogenic emissions of carbon dioxide (CO_2) to their 1990 level by 2000. The Strategy stated that 'New Zealand retains its carbon dioxide emissions to 20 per cent below their 1990 levels, subject to certain conditions, including cost-effectiveness, not reducing our competitive advantage in international trade, and having a net benefit to New Zealand Society'.[6]

A/NZ's ETS Pathway: Global Compliance

In 1992 New Zealand became the 34th nation to sign the United Nations Framework Convention for Climate Change (UNFCCC; The Convention) at the 1992 United Nations Conference on Environment and Development, in Rio de Janeiro. This signalled New Zealand's commitment to reduce GHG emissions in line with global responses to climate change. The convention required Governments to 'share information on emissions and policies, to launch national strategies dealing with GHG emissions, to provide financial and technological support to developing countries and to cooperate in preparing for the impacts of climate change'.[7]

New Zealand ratified The Convention in 1993 and significantly adjusted previous targets and government policies to bring these into line with those of other Annex 1 countries (Industrialised Countries).[8] The new target for lowering GHG was now set at 1990 levels by 2020. This was significantly less than the previous reduction target of 20 per cent below 1990 levels by 2020.

The key statement in New Zealand's first *National Communication* (1994) noted that New Zealand would reduce the emissions of the most significant GHG (carbon dioxide) to 1990 levels by the year 2000, which was to be accomplished through a mix of policies and legislation. The *National Communication* indicated that local bodies would take responsibility for adaptation and at present these measures 'have been concentrated on coastal policy and natural hazard mitigation'.[9]

In 1997 New Zealand became one of the industrialised nations to sign the UNFCCC Kyoto Protocol. It sets 'legally binding reduction commitments for 6 green house gasses'.[10] These GHG are produced by Annex 1 countries and are carbon dioxide, methane, nitrous oxide, sulphur hexafluoride, hydrofluorocarbons, and perfluorocarbons.[11] New Zealand ratified the Kyoto Protocol in 2002 and it became operative in 2005.

A/NZ's ETS Pathway: National Compliance

The Climate Change Response Act, 2002 was enacted to help New Zealand to meet its obligations under the Kyoto Protocol. The Ministry for the Environment became the inventory agency and was responsible for 'the overall development, compilation and submission of the inventory to the Convention secretariat'.[12] Between 2006 and 2009 New Zealand reiterated its two national targets for reducing New Zealand's GHG emissions. These were communicated to The Convention in the Fifth *National Communication*: A medium term responsibility target of 10–20 per cent GHG reduction below 1990 levels by 2020; and a long-term target of 50 per cent reduction in net GHG from 1990 levels by 2050.[13]

A/NZ expected to meet its targets through domestic emission reductions, CCS in forests, and by purchasing emission reduction units from other partner countries.[14] These would be possible through land-use changes and forestry and an active participation in global carbon markets.

NZETS As a Mitigation Strategy

The NZETS was introduced through the Climate Change Response (Emissions Trading) Amendment Act, 2008. It was then amended in November 2009 and again in November 2012. The primary purpose of the Act is to allow New Zealand to meet the agreed-to criteria for reduced GHG emissions as a signatory to the Kyoto Protocol and the UNFCCC. The NZETS was the preferred option over and above a carbon tax, which had been a preferred mechanism by environmental groups including the New Zealand Green Party. The New Zealand Government did not favour a carbon tax and the ETS as passed under the legislation did not cap the amount of allowable GHG emissions—a major criticism of the 2008 Act. Instead the NZETS 'prices but does not limit, New Zealand's emissions of six gases (outside of agriculture): CO_2, CH_4, N_2O, HFCs, perfluorocarbons, and Sulfur Hexafluoride (SF_6)'.[15] In his working paper on the effectiveness of the NZETS David Bullock explains that a carbon tax acts as an incentive to reduce the harmful GHG and consequently is the set tax each producer must pay, 'the person who creates the harm should pay for it'.[16]

Emissions Trading Schemes, New Zealand's excluded, are generally cap-and-trade schemes. A government will set a cap on the maximum percentage of GHG emissions per annum. Tradable units or carbon credits offset extra emissions over and above the cap. A predetermined acceptable level of GHG is set by the Government 'who then allocates, by sale or free, permits that allow a holder to emit a proportion of that total amount'.[17,18] The permits are tradable to other firms so that any excess units can be sold on. For industry members who cannot reduce emissions adequately, they can buy permits to help account for their emissions.[19] The New Zealand government chose not to place a cap on the emissions so the NZETS is a carbon trading scheme with no set emissions limit. Emissions Trading Schemes are economically focused so are often at the expense of social, environmental, and cultural factors. The key difference between a carbon tax and a cap-in-trade scheme is that unlike a carbon tax that encourages reduction in GHG emissions, the cap-in-trade schemes do not.

How It Works: Carbon Credits and Allocations

The Environmental Protection Agency (EPA) manages the administration of the NZETS and ensures 'compliance with the scheme and provides reports and market information... [It] also operate the New Zealand Emissions Unit Register, where transactions take place'.[20] The NZETS relies on the allocation of GHG units, which are given to participating enterprises. An emissions unit can represent either one metric tonne of carbon dioxide, or the equivalent of any other GHG. There are two types of units allowed under the scheme. The New Zealand Government Units (NZU) also known as the domestic NZ Units are a fixed amount. The NZU are allocated freely to each eligible GHG enterprise. The second category of unit is the international Kyoto-compliant unit. There is no limit on how many of these may be imported by eligible GHG enterprises, and can be made up from various categories of Kyoto-compliant units. All the eligible industries under the NZETS scheme surrender units (both NZU and Kyoto-compliance units) at a prescribed rate determined by the production output of each enterprise. The units are allocated by the New Zealand Government (NZUs) or can be traded either on the international market or among participants in the NZETS. They are commonly referred to as carbon credits. The amount of GHG emissions generated by each eligible GHG enterprise is not capped within New Zealand (Table 5.1).[21]

Table 5.1   New Zealand's GHG units

Forestry New Zealand Units (NZUs)	These are NZUs given to foresters in the ETS. They may be converted to NZ Assigned Amount Units (AAUs) for offshore sale
Other NZUs	These are all other NZUs, including those given to industrial allocation recipients. They cannot be converted to NZ AAUs for offshore sale
AAUs	New Zealand based AAUs can be either: Forestry NZUs that have been converted into NZ AAUs; or NZ AAUs that have been granted to companies in New Zealand that have participated in the Projects to Reduce Emissions (PRE) or the Permanent Sink Initiative (PFSI)
Certified Emissions Reduction Units (CERs)	CERs are units generated by Clean Development Mechanism (CDM) projects offshore. These units are able to be purchased by participants in the ETS and used to meet their surrender obligations
Emissions Reduction Units (EMRs)	ERUs are units generated by Joint Implementation (JI) projects. These units can be purchased by participants in the ETS and used to meet their surrender obligations
Removal Units (RMUs)	RMUs are Kyoto Protocol units generated through storing carbon in trees. These units can be purchased by participants in the NZETS and used to meet their surrender obligations
NZ$25 fixed price option	Companies have the option to pay the government a NZ$25 fixed price per unit, rather than surrender the other types of eligible units

Source: New Zealand Environment Protection Agency, *The New Zealand Emissions Trading Scheme. ETS 2013 – Facts and Figures.* (http://www.epa.govt.nz/e-m-t/pages/about-ets.aspx). Last retrieved 15 January 2015

Units are allocated according to whether or not the industry participant is a moderate (producing more than 800 tonnes of carbon dioxide equivalent emissions per million dollars of revenue), or high emissions intensive industry (producing 1600 tonnes per million dollars of revenue).[22] The key eligible GHG enterprises in the NZETS are forestry (pre-1990; post-1989), stationary energy, industrial processes, liquid fossil fuels, agriculture, waste and other sectors.

First Changes to the NZETS

In 2009 the government passed the Climate Change Response (Moderated Emissions Trading) Act. This immediately met with widespread criticism both nationally and internationally over two main areas of concern: the high social cost and the lack of real incentive to reduce GHG emissions.

High social costs are found in the government's NZUs allocation process which means that effectively the New Zealand taxpayer will be paying for the NZETS through higher household costs. The costs will be, in the words of the Minister for Climate Change Issues (2009), 'very significant' and expected to be in the region of $NZ900 million by 2030.[23] The taxpayer will bear the cost of higher priced consumer products that rely on manufacturing processes from ETs eligible industries, such as higher electricity prices, food, and clothing goods.

Secondly NZETS does not actively encourage the reduction of GHG emissions. This has come about because there is no cap within New Zealand on the amount of GHG emissions that each emission intensive industry produces. The highly internationalised unit-trading scheme coupled with the free allocation of NZU allows each industry member to factor in the cost of GHG emissions on both short- and long-term economic forecasting. The amount of units they surrender under the scheme is proportional to their units of production. Even if the company were to use all of their freely allocated NZGU, they can buy in any shortfall in Kyoto-compliant Units. This means the NZETS is controlled to some extent by an international trading market, which further entrench economic factors as a driving force and determinant of the reduction in GHG emissions. Thus, there are no socially or environmentally responsible incentives to reduce GHG, at least not at the expense of profits and economic growth.

THE 2012 AMENDMENT

The Act was amended again in 2012. The key changes were to extend the transitional measure to reduce the cost and extend the ability of participants to meet their emissions obligations under the scheme. The second change was to defer the date that the agriculture industry entered the scheme. The agriculture sector was originally scheduled to come into the Scheme on 1 January 2015, but intense lobbying from industry representative, Federated Farmers, meant that the inclusion of this sector has been highly contested from the start despite the industry's significant contribution to GHG emissions. Currently the agriculture sector must report emissions, but does not have to surrender any units in payment for GHG production. As such the agriculture sector is being heavily subsidised for its non-participation in the scheme. In 2017 the industry was still not part of the NZETS, though producers continue to monitor their emissions.

The third and fourth key changes in 2012 allowed the government to introduce more measures to encourage reforestation, and gave extra powers for the government to increase the supply of NZUs.

The government's reason for the changes was that 'now is not the right time, in an uncertain economic environment, to put more costs on household and businesses'.[24] The Act will be reviewed again in 2015 and will include 'consideration of progress made in completing a new comprehensive international agreement',[25] suggesting that New Zealand's efforts to reduce GHG emissions continues to be led by the international sector and is market driven, as opposed to environmentally and socially driven.

THE NZETS AND MĀORI

Climate change will have an impact across all of the A/NZ economy, social, and environmental sectors including that of Māori. Māori Iwi (tribal groups), whānau (family groups), and hapū (sub-tribes) have status as the first people of A/NZ that was officially recognised with the signing of the Treaty of Waitangi between some Māori tribes and the British Crown 175 years ago on 6 February 1840.[26,27] What followed for Māori was 135 years of land loss, assimilation, and political marginalisation. In 1975 the Treaty of Waitangi Act installed the Waitangi Tribunal, which was a Commission of Enquiry set up to investigate tribal grievances from 1975 forwards. The Act was amended in 1985 to allow retrospective grievances back to 1840. The Treaty of Waitangi Settlements negotiated between Iwi and the Crown have begun the reinstatement of Māori tribal groups as the recognised treaty partners. The Waitangi Treaty Settlements are the long-awaited compensation for breaches of the Treaty of Waitangi. The role of Treaty Settlements and in particular Māori tribal development in a post-settlement era is important for two reasons. For many tribal groups a Treaty settlement allows them to reassert mana (recognised power and authority) over their specific regions, resources, and the people that make up their whakapapa. Secondly, the bulk of a settlement is usually monetary compensation, thus providing an economic base to begin economic development for the future. At best the settlements provide the money and other assets to begin social and cultural development, thus building social and cultural confidence.[28] Growing an economic base leads to overall well-being of an indigenous group.[29] The Treaty Settlement process also allows for a new era of engagement and relationship with the Crown (the New Zealand Government) that focuses on future and intergenerational development along parallel pathways to that of non-Māori New Zealanders. That said, political marginalisation remains a major factor to tribal self-determination over their territories and resources.

But Māori through their economic growth and development in areas targeted for reducing GHG are also part of the problem. Inclusion into the NZETS will inevitably create tensions for Māori groups in realising effective measures that meet all the environmental, cultural, social, and economic aspirations for intergenerational growth.

The Māori economy is a major part of the overall A/NZ economy. In a report that reviewed the developing Māori economy, Carter et al. noted in 2011 that the 'Māori economy is dominated by Māori Small to Medium Enterprises (SMEs), Māori Trust/Incorporations, and Rūnanga (Collective assets)'.[30] They also noted the Social Accounting Matrix (SAM) for the Māori economy, which 'models household and individual data, including labour force statistics, household incomes and net savings', was measured against the 'rest of New Zealand'[31] to provide a picture of the Māori economy both nationally and regionally. Recent research has investigated the diverse nature of the many enterprises that make up the Māori economy, particularly the hidden factors of economic production. Many of these revolve around the principles of MEK.

All Iwi and Māori eligible GHG enterprises fit within the NZETS. The impact from a scheme such as the NZETS will be significant for Māori because it is outside of any MEK approach that seeks balance across the environment, social, cultural, and economic factors. The majority of Māori assets are land-based (agriculture, geothermal power production, horticulture) and Iwi are also heavily involved in the fishing industry and tourism. Under MEK climate change impact on the economic factors will ultimately impact on the social, cultural, and environmental factors across New Zealand and in each tribal region. The Māori economic enterprises though are included in the NZETS, which is primarily economically driven.

In 2008 a report outlining the impacts from the NZETS on Māori was released. The Cognitas Report was designed to provide 'an analysis of the likely impacts of the government's proposed emissions trading scheme on Māori' and considered how it might affect Māori differently from non-Māori in both positive and negative contexts.[32] The report investigated the impact on Māori households through to the agriculture and fishing industries; how the ETS might affect geothermal electricity generation (as many of the fields sit under Māori-owned land), and the power generation enterprises that are owned and controlled by Māori.

There were several assumptions made within the report that contextualised the research. Key to this was the assumption that Māori interests lie in cultural, environmental, and economic values of sustainability. The value of sustainability was further detailed through discussion on economic sustainability, land sustainability, and cultural sustainability. The report also outlined the important and central place of land to Māori cultural values, beliefs, and identity and that Māori land is 'Taonga tuku iho' (a treasure passed from generation to generation).[33] The authors commented 'it is not the mere passing of title that is significant, rather it is the network of turangawaewae links that the land ownership represents that are to be maintained'[34] across generations. Many of the climate change practices currently being enacted relate specifically to this belief that Māori are kaitiaki of the land and resources and are here to leave everything in a better condition for the future generations. The cultural ties Māori have with their tribal regions and the 'greater interest [Māori] have in preserving the economic viability of their assets within those regions' assumes that '[Māori]…are concerned with sustaining their communities within those regions, and also the environment in which those communities live and operate'.[35] Thus the knowledge 'that Māori aren't going anywhere'[36] means that whatever choices and decisions each group makes in the future will be couched in the tikaka that best serve the cultural landscape and its people in question.

Other assumed interests were self-government of Māori assets—by Māori, for Māori. The assumption here is that Māori know the best way to develop their assets and in ways that benefit them culturally and economically. A further important assumption was that Māori are not interested in bearing a heavier responsibility to that of non-Māori, but one that is proportionate to their relative contribution to GHG emissions. Also, the level of burden should be 'concerned with their level of economic development relative to non-Māori (as a consequence of past Crown actions or otherwise)'.[37] The report also identifies Māori interest in the international emissions trading and the economic opportunities that trading offers.

The Cognitas Report detailed the NZETS impact on different sectors of the Māori economy with a warning that 'determining where the overall Māori preference lies in relation to the ETS is both mis-directed and impossible given the available data…there is no unitary Māori economy [so] it is not possible to discern a single Māori preference regarding the different features of the proposed ETS'.[38] The Report's authors added that 'without knowing the precise mix of land and other interest of particular Māori organisations and individuals it is not possible to determine whether the ETS on balance helps or harms their net interests'.[39] Regardless they were able to make some predictions of the impact on several economic areas. At the centre of these was the level of social impact on Māori households. This included future fuel price rises and electricity costs that will have a greater impact on low-socioeconomic Māori households. Future employment will also be a factor because of the prevalence of Māori employed in the primary industries (e.g. fishing, meat production) and industries reliant on high levels of fossil fuel usage. The forestry sector should increase potential employment through the carbon sink incentives in the NZETS.[40] The Report also outlined the relatively positive economic potential for Māori in the forestry, fishing, agricultural, and geothermal industries—many of which make up economic enterprises on tribal Māori lands.

The Report stressed the 'very real need and urgency around assembling detailed inventory and mapping of Māori forest land ownership and published data on Māori and non-Māori interests in pre-1990 and post-1989 forests to enable development options for Māori'.[41] The authors also raised the issues of cultural use of Māori land, such as the development of papakainga housing (village and/or community housing) in terms of changing land-use; and the local government considerations around land-use changes when considering future adaptation measures. The environmental and cultural factors were not dealt with in the report, but these will affect all the areas surveyed. For example most of the Māori-owned geothermal enterprises are owned by whānau trusts and larger whānau incorporations. These are governed and managed using Māori knowledge frameworks, practices, and values. The overall objective is to provide whānau owners with the means for intergenerational social and economic development. These objectives are underpinned by their whakapapa organisational structure and its guiding principles of manaakitaka, whakawhanaukataka, and kaitiakitaka, remain in tension with the economic focus of the NZETS.

One part of the NZETS is the CCS potential of reforested land. Māori organisations own large acreages of land that are suitable for reforestation and the associated benefits from increasing carbon sinks as part of the NZETS. In a study undertaken on the North Island's east coast (Tairawhiti), Garth Harmsworth argued that while Māori are generally supportive of reforesting land, they were concerned 'that government policies are not aligning with broader Māori aspirations and objectives', and that policies that do include 'Māori aspirations, cultural values and rights under the Treaty of Waitangi'[42] would be beneficial to A/NZ and help achieve the Kyoto goals. Māori concerns included the retention of Māori land ownership and control, costs associated with joining an ETS scheme, long-terms costs such as rates on reforested land, and limits to possible income from lands that must now remain in a forested state. Harmsworth identified several key areas for consideration if Māori land-owners were to consider reforestation projects. Firstly it was important to identify exactly what was 'Kyoto forest' that would be suited for carbon trading and if this included large areas currently covered in beech, kānuka, rata, broadleaf, and other native species. Secondly that the different land governance and management structures (as per the Te Ture Whenua Māori Land Act definitions) were recognised, and thirdly that future aspirations (economic, social, and cultural) would be recognised and unhindered if land became part of the NZETS.[43]

Another initiative currently underway investigates climate resilient Māori land investment decisions with Ngāti Porou Iwi in the Waiapu catchment area of North Island's east coast. The Waiapu catchment and in particular the Waiapu river feature strongly in Ngāti Porou whakapapa and as such the Iwi are an integral part of the river and surrounding environment and ecosystems. One of the leading causes for concern is increased soil erosion and sediment levels that are changing the land and waterway conditions. The predicted climate change impacts include increased rainfall events and more extreme droughts that will potentially further change the river and its environs. A LandCare New Zealand research team worked with the Iwi to find ways to mitigate the risks by assessing the economic and Māori values impacts of different land-use decisions within a number of climate change scenarios. Alternative land-use included reforestation and horticulture. Each scenario was assessed on its ability to meet intergenerational development and growth, Ngāti Porou values, and landowner aspirations. The Iwi indicated that any decisions made were not to impact negatively on Ngāti Porou well-being and that the values, beliefs, and practices would underpin the future actions taken.[44]

As Bullock has noted, 'The NZETS hopefully signals a new era of climate change policy in New Zealand.'[45] There are still concerns about the efficiency of the scheme to reduce significantly New Zealand's GHG emissions within the set time period. The economic focus also presents difficulties for environmental sustainability, as it does not adequately ensure that GHG emissions will be voluntarily reduced at a rate that will meet the proposed reductions by 2030. As a member of the Pacific Islands Forum, New Zealand has been party to the Majuro Declaration. The Declaration is designed to 'highlight the commitment of the leaders of the Pacific Island Forum countries to the reduction and phasing down of greenhouse gas pollution worldwide, with the leaders wanting to "spark a new wave of climate leadership"'. New Zealand currently is among the worst of the Annex 1 countries in meeting their predicted reductions. NZETS detractors state that 'New Zealand does not have the luxury' to exclude agriculture from the mix if 'the scheme is going to create real and overseas comparable emissions reductions'.[46] The future of the overall New Zealand economy weighs heavily in any decision to include Agricultural emissions. A/NZ must meet her responsibilities and obligations to the relationships with her Pacific Islands Forum partners yet so far her key mitigation strategy is not producing the expected results. The NZETS has produced a 29 per cent decline in GHG removals since 1990, which is well below the expected targets, and seen as evidence that New Zealand's scheme is not working. The environmental sector call it an 'ETS scheme you have when you are not having an ETS', while the government and big business sector applaud it for the economic freedom that it provides.

Māori are interested in changing land-use for their lands, but only if it does not compromise their cultural values, and remains within a MM-based set of guidelines and practices. It seems then that any further progression of Kyoto Protocols will need to ensure that A/NZ develops 'culturally based carbon contract models'[47] that meet the four Treaty of Waitangi principles of partnership, protection, participation, and potential.

NOTES

1. Bullock, 2009, 2.
2. Bullock, 2009, 2.
3. The New Zealand Herald – *NZ's Emissions Impossible*, 16 December 2014. Last retrieved accessed 15 January 2015.
4. Tim Grosser in Climate Change: *NZ's Emission Impossible*, in The New Zealand Herald, 16 December, 2014. (http://www.nzherald.co.nz/news/print.cfm?pbjectid=11374647]) Last retrieved 15 January 2015.
5. (www.mfe.govt.nz/publications/climate/nz-fifth-national-communication/page2.html) Last accessed 28/8/2014.

6. (www.mfe.govt.nz/publications/climate/nz-fifth-national-communication/page2.html) Last accessed 28/8/2014.
7. MfE report 2005, cited in Bullock, 2009, 6.
8. Bullock, 2009, 6.
9. www.mfe.govt.nz/publications/climate/nz-fifth-national-communication/page2.html Executive Summary p. 6. Last accessed 28/8/2014.
10. Bullock, 2009, 6.
11. Bullock, 2009, 6.
12. www.mfe.govt.nz/publications/climate/nz-fifth-national-communication/page2.html (last accessed 28/8/2014).
13. www.mfe.govt.nz/publications/climate/nz-fifth-national-communication/page2.html Executive Summary, 6 (last accessed 28/8/2014).
14. www.mfe.govt.nz/publications/climate/nz-fifth-national-communication/page2.html Executive Summary, 1.4 *Policies and Measures*, 6 (last accessed 28/8/2014).
15. Hopkins et al., 2015.
16. Bullock, 2009, 10.
17. Bullock, 2009, 11.
18. Hopkins et al., 2015.
19. Bullock, 2009, 11.
20. Environment Protection Agency, *About the ETS*. (http://www.epa.govt.nz/e-m-t/Pages/About-ets.aspx). Last accessed 15 January 2015.
21. For more detail on the mechanics of the NZETS refer to Bullock, D. (2009); Environmental Protection Agency, *ETS 2013 – Facts and Figures* (http://www.epa.govt.nz/e-m-t/pages/about-ets.aspx); New Zealand's Ministry for the Environment, (http://www.mfe.govt.nz).
22. www.epa.govt.nz/publications/EPA_He_Whetu_Marama.PDF.
23. Dr Nick Smith, Minister for Climate Change Issues, 9 October 2009. Amendments for a Moderated NZ ETS and Second Order amendments to the Climate Change Response Act, in *Cabinet Paper EGI (09) 13/2*. Office of the Minister for Climate Change Issues. http://www.mfe.govt.nz/cabinet-papers/egi-09-132.html.
24. Ministry for the Environment, *2012 Amendments to the New Zealand Emissions Trading Scheme (NZ ETS): Questions and answers*. Last accessed 21 January 2015, 2.
25. Ministry for the Environment, *Amendments to the New Zealand Emissions Trading Schemes 2012*, 5.
26. Orange, 1987.
27. Kawharu, 1989.
28. Carter et al., 2011, 17.
29. Carter et al., 2011, 17.
30. Carter et al., 2011, 27.
31. Carter et al., 2011, 27.

32. Insley and Meade, 2008.
33. Insley and Meade, 2008, 2.
34. Insley and Meade, 2008, 2.
35. Insley and Meade, 2008, 2.
36. Personal statement, Sir Mark Solomon, Kaiwhakahaere of Te Runanga o Ngai Tahu (1997–2017).
37. Insley and Meade, 2008, 3.
38. Insley and Meade, 2008, 3.
39. Insley and Meade, 2008, 49.
40. Insley and Meade, 2008.
41. Insley and Meade, 2008.
42. Harmsworth, 2003.
43. Harmsworth, 2003, 5.1.
44. Personal conversation. Shaun Awatere presented the research and findings at the Pacific Climate Change Conference, Wellington, February 2018, and the information given at the conference is included here with permission.
45. Bullock, 2009, 41.
46. Bullock, 2009, 41.
47. LandCare New Zealand Report, *Forests and shrublands as a place to store carbon.*

References

Bullock, D. (2009). *The New Zealand Emissions Trading Scheme: A Step in the Right Direction?* (Institute of Policy Studies Working Paper 09/04, March 2009). Wellington: School of Government Studies, University of Victoria.

Carter, L., Kamou, R., & Barrett, M. (2011). *Literature Review and Programme Report. Te Pae Tawhiti Maori Economic Development Porgramme.* Published Report for Nga Pae o Te Maramatanga, University of Auckland.

Harmsworth, G. (2003). *Maori Perspectives on Kyoto Policy: Interim Results. Reducing Greenhouse Gas Emissions from the Terrestrial Biosphere (C09X0212).* Discussion Paper for Policy Agencies (Climate Change Office; MfE, MAF, TPK). Palmerston North: Landcare Research NZ.

Hopkins, D., Campbell-Hunt, C., Carter, L., Higham, J., & Rosin, C. (2015). Climate Change and Aotearoa New Zealand. *WIRes Climate Change.* https://doi.org/10.1002/wcc.355.

Insley, C., & Meade, R. (2008). *Māori Impacts from the Emissions Trading Scheme. Detailed Analysis and Conclusions.* Wellington: Ministry for the Environment. Prepared by 37 Degrees South and The Cognitas Advisory Services.

Kawharu, I. H. (Ed.). (1989). *Waitangi. Maori and Pakeha Perspectives of the Treaty of Waitangi.* Oxford: Oxford University Press.

Orange, C. (1987). *The Treaty of Waitangi.* Wellington: Allen and Unwen/Port Nicholson Press.

CHAPTER 6

Aotearoa/New Zealand Adaptation Strategies and Practices

Abstract IPCC has criticised Aotearoa/New Zealand for not being proactive with adaptation measures to date. However, some initiatives have been developed and actioned mainly by Māori and wider community groups. Local government has been given the mandate from central government to lead the way in climate change adaption and this chapter outlines some of the initiatives and challenges towards effective adaptation. A case study from the South Island hapū, Kāti Huirapa is one example of a Māori-led initiative to providing effective and culturally responsive actions for future sustainability of vulnerable resources. This chapter promotes a Mātauraka-a-Iwi approach to climate change that best illustrates the idea of states of transition and renegotiating relationships in order to move forward.

Keywords Adaptation • Local government • Community strategies • Kāti Huirapa adaptation project

Adaptation differs from mitigation in that it is the area where actions to reduce the impact from climate change takes place. Adaptation refers to 'actions targeted at a specific vulnerable system, in response to actual or expected climate change, with the objective to either limit negative impacts or exploit positive ones'.[1] It involves establishing the vulnerability of the systems to any changes, and determining the risk to ensure appropriate

L. Carter, *Indigenous Pacific Approaches to Climate Change*,
Palgrave Studies in Disaster Anthropology,
https://doi.org/10.1007/978-3-319-96439-3_6

long-term management responses can be put in place. Should a system reach its tipping point—that is the point at which the current management strategy will no longer meet the objectives for successfully managing the system—the response may fail.[2] Kwadijk et al. argue that successful adaptation involves dealing with the predictability of climate change (such as sea-level rise); non-climatic conditions (current and future utilisation of the specific system); and the time horizon (short- or long-term actions).[3] In the context of IK and adaptation, methods chosen are 'strategies that enable the individual or the community to cope with or adjust to the impacts of climate in local areas' as it allows for strategies that connect people directly to their environments through the relationships formed over time. This includes the changes that occur within the environment that are both human-induced or part of the natural cycle of ecological adjustment.[4]

In the context of climate change in A/NZ, adaptation strategies are linked to natural hazards management and sustainable development. Natural hazard management is the responsibility of local and central government and also Iwi/Māori whose territories include the coastal, marine, and land environments and all interactions and interrelated relationships amongst them. Therefore, they are inextricably linked into the whakapapa of Māori communities and the kaitiaki role that Māori have within their respective takiwā (territories). The established Iwi relationships and treaty-based partnerships with government and LCs become increasingly important when thinking about climate change adaptation in A/NZ. As with Māori responsibilities for mitigation policies and actions, Māori adaptation responsibilities involve creating balanced processes and solutions across the economic, social, cultural, and environmental context. These are invariably informed through Mātauraka-a-Iwi that underpin all Iwi interactions within their respective takiwā. The concept of co-production of knowledge also plays an important part in adaptation because Iwi knowledge can be reinforced with non-Iwi knowledge to ensure the most beneficial outcomes for all those who interact within Iwi territories. An important part of adaptation is dealing with the predictability of climate change.[5] In order to do this communities and Iwi need to know what the expected changes of the current state are expected to be. Once the system changes (states of transition) are known then planning for future management will incorporate vulnerability, adaptive change, and indicate future capacity needs. One such plan that identifies the system changes is the 2013 NIWA report for Arowhenua Pā in south Canterbury, New Zealand. The key findings from this report emphasised a TEK approach to future adaptation strategies for the Arowhenua Pā community.

AROWHENUA PĀ AND THE NIWA REPORT

Te Rūnanga o Arowhenua is within the Kāti Huirapa takiwā and is one of 18 regional councils from the wider Kāi Tahu iwi grouping. It is situated near the South Island town of Te Umu Kaha (Temuka) and sits between the Opihi and Te Umu Kaha rivers. The New Zealand National Institute for Weather and Atmospheric Research Ltd. (NIWA) conducted research on the flooding hazard within the Arowhenua community to establish what had occurred in the past, and what was happening now following un-natural modification of the river and wetlands systems. This information was then used to determine how this would affect the Pā in the future with climate change as a new mitigating factor.[6] Although it is not the intention of the author to recreate the report here, it serves as an example establishing the guiding factors to be considered from the Arowhenua community's role as kaitiaki of this region. The responsibilities will be similar to what other Māori coastal communities will need to consider when adapting to climate change as an additional challenge in current and future kaitiaki obligations. The long-term planning strategies will need to take into account the long-term impact of climate change. Therefore any adaptation measures will need to be culturally, economically, environmentally, and socially relevant for Māori to achieve sustained, intergenerational development and growth. The Kāi Tahu vision statement, *Mō mātou, ā, mō ngā uri ā muri nei*, (for us and our generations still to come) emphasises the longevity of the Iwi and the future intergenerational considerations that underpin all current strategies and actions. The same sentiments can be found in most Iwi strategic plans as well as the underpinning values of whakapapa, kaitiakitaka, whanaukataka, and manaakitaka.

One key finding from the NIWA Arowhenua Pā research was developing future provisions for building capacity and capability among whānau members. The social capital aspect of future development and adaptation is particularly poignant because of the large numbers of Iwi members who live away from tribal areas. This will become more challenging as young members move away in search of employment and better quality of life. Increases in the inability of the territory to sustain members may be exacerbated through challenges such as climate change and echo those of Pacific Island nations. Climate change 'refugees' could become a localised term for forced evacuation of A/NZ coastal communities, as with our Pacific neighbours.

The NIWA research study identified four areas that will require consideration for any successful future hazard management (flooding in particular) in the Arowhenua Pā community. These are highlighted as the strengths in the following table. Also highlighted are the risks and challenges to effective, culturally relevant adaptation measures being put in place. Both the strengths and the risks will be similar for all Māori/Iwi communities across A/NZ and need to be mitigated in culturally relevant ways for effective long-term adaptation measures to be put in place (Table 6.1).

Table 6.1 Principal determinants of Māori community sensitivity and adaptive capacity

Strengths	Governed through cultural values and approaches:
Social networks, conventions, and transformation,	Whakapapa
Knowledge skills and expertise,	Wairua
Resourcing and finance,	Tikanga
Institutions, governance, and policy,	Kawa
	Whanaungatanga
	Manaakitanga
	Kotahitanga
	Mahinga kai (knowledge of place, environment, and resources)
	Resilience
Risks and challenges	Decreasing cultural strengths,
	Rapid transformations in Māori community structures,
	Population migration,
	Decreased Māori land holdings and ownership,
	Capacity to deal with environmental risks,
	Lack of financing,
	Non-Māori resource management regimes,
	Land-use changes

Source: NIWA/Taihoko Nukurangi, Report *Maori Community Adaptation to Climate Variability and Change. Examining risk, vulnerability and adaptive strategies with Ngati Huirapa at Arowhenua Pa, Te Umu Kaha (Temuka), New Zealand.* Prepared for Te Runanga o Arowhenua Society Incorporated and the New Zealand Climate Change Research Institute – Victoria University, Wellington, March 2012, p. 106

The various strengths and associated cultural practices are examples of attributes found in a generalised MEK framework. The report provides information about system changes that will enable Arowhenua to understand the vulnerability of their environment and how to meet these head-on. The NIWA report, while dealing specifically with Arowhenua Pā in South Canterbury, has relevance to other coastal Māori communities to guide towards positive changes and solutions.

In 2016 TRoNT engaged NIWA to produce a set of 18 reports for each of the tribal rūnaka territories. These capture the region-specific challenges each Rūnaka is likely to face. The reports are useful in helping planning and strategies for meeting the challenges head-on. Utilising MEK frameworks will mean that Kāi Tahu rūnaka will take the lead in planning and executing adaptation strategies within their respective takiwā. Reports such as the NIWA ones for Kāi Tahu will enable the other two main groups responsible for overseeing any A/NZ adaptation policies and actions to ensure that their solutions take full cognisance of Iwi knowledge, values, and intergenerational aspirations. The treaty partners involved with Iwi within the tribal territories are local and central government agencies.

Adaptation: Local and Central Government

The other main areas of responsibility for adaptation are assigned to the Central New Zealand Government and LCs. Both of these levels of governance and management have partnership relationships with Iwi and therefore have responsibilities to ensure Iwi aspirations and goals are included in any adaptation policies, legislation, and actions. The first is central government whose role is to provide policies and the legislative framework and governance 'to support local councils and communities in how they adapt to climate change'.[7] The second group is local government (councils: local and regional) whose responsibility it is to prepare for and manage the risks 'because they are best placed to know what is appropriate for their region'.[8] The legislative framework, policies, and strategies put in place by these two groups will allow industry and business, communities, and individuals to undertake practical climate change adaptation measures in their respective communities.[9]

LOCAL GOVERNMENT RESPONSIBILITIES

The Ministry for the Environment has a large number of guidelines documents for local government particularly local councils (LCs). These include tool kits and examples of case studies that various LCs and regional bodies have undertaken. One particularly useful document is the *Climate Change and Long-Term Council Community Planning* document, which sets out the key guiding principles to planning for climate change. These are the future generational needs, sustainability, the ethic of stewardship/ kaitiakitaka, consultation and participation, and financial responsibility. The guidelines also ask LCs to state if policies are to avoid, remedy, or mitigate any adverse effects, and how measures undertaken will produce resilient communities.[10]

As with the MEK ethic, the principles take into account the four key factors towards sustainable and intergenerational development: economic, environmental, cultural, and social. The LCs are reminded that adaptation is about taking steps to minimise threats in order to maximise opportunities resulting from climate change.[11] The document instructs LCs to remember that 'a key element in adapting to climate change will be identifying flexible options that allow an incremental response over time' and 'identifying options that help reduce current vulnerability.'[12] There is useful information on exactly how the climate might change and some of the ways in which LCs have already initiated adaptation measures across New Zealand.

LCs are also NZETS participants through their involvement in waste management, forestry, and in some cases agriculture. In the case of waste management for example, a Ministry for the Environment (MfE) document states that as there is no free allocation of GHG units for waste management, 'it is expected that the costs of purchasing compliance units will be passed onto the users of the landfill. Consequently the costs of operating a landfill and of disposing waste will increase'.[13] In some cases LCs face a future with an aging population and consequently a shrinking rates base—Invercargill City Council for example. The added costs and shrinking rates base will mean that new ways for funding will need to be sought. Large cities, such as Auckland face challenges to infrastructure, and technological changes needed to manage the future growth and development in a climate change era. It means that councils will have to think smarter about how to initiate strategies and policies so that the GHG emissions from council-owned and operated industries will be more environmentally friendly and cost-effective for ratepayers. This places a huge burden on current councils to provide workable and sustainable strategies in their Long-Term Council Community Plans (LTCCPs).

Many LCs have already introduced climate change into their long-term and short-term planning strategies. These comply with the mitigation aspects of the government's role such as the Waste Minimisation Act, 2008. Dunedin City Council (DCC) for example has its waste minimisation strategy which calls for 'Reduce (waste generation); Re-use (further use of the existing product); Recycle (reprocessing waste products)'.[14] The DCC promotes 'Recovery (extraction of material or energy for further use or processing; composting); Treatment (treating hazardous wastes for environment-friendly disposal); and Disposal (waste deposit sites/land-fills).[15] DCC, as do all city councils, has strategies and planning built into its LTCCP for areas such as energy, transport infrastructure, and land-use changes. These identify any future considerations towards the impact from climate change. Many council documents outline the opportunity for innovative solutions to areas such as transport infrastructure (cycle ways, carpooling), and the potential risks through risk assessment and operational planning and processes.

As with Māori tribal and whānau enterprises, the LCs are responsible for mitigation (the NZETS) and adaptation responses that are often in tension with their economic activities. This also guides to some extent Iwi and LC social responsibilities, which includes engagement with the treaty partners with whom they share territorial regions. Where there is a local response to adaptation, led and implemented by the local communities, it is usually a community initiative. As ratepayers and the treaty partner with LCs, Iwi are also an integral part of local solutions.

The IPCC assessed two key climate change challenges for New Zealand: sea-level rise and flooding. In A/NZ, the coastal areas can be low lying and part of river flood plain systems. This makes them susceptible to flooding which causes further erosion. The problem also occurs along riverbanks where extensive environmental modification has taken place, not only in rural and semi-rural environments, but also in urban and industrial areas. Activities such as housing and industrial development have modified some stream and rivers in cities and towns to the extent that the water quality and damage to the flora and fauna have suffered. DCC for example faces a major future headache for much of South Dunedin. This area is made up of a mixture of industrial, schools, parks and residential housing that covers reclaimed wetlands and sits behind a highly modified sand-dune area.

In 2016 New Zealand's Environment Commissioner listed South Dunedin as a priority area for climate change action. The Dunedin mayor Dave Cull was quoted as saying that 'the impending inundation of South Dunedin is one of the biggest issues confronting the city over the next couple of decades'.[16] The Mayor stops short of signalling mass relocation and insists that steps such as raised housing and strengthening the drainage and waste-water facilities will be a more cautious and sustainable approach. It is probable that a relocation scenario may be a challenge for future Dunedin councils. Both central and local government, industry, communities, and Iwi are taking the threat seriously, and are working to future proof the living and built environments.

Other local initiatives that involve all the community are dune management and restoration. A/NZ as a group of Pacific islands is subject to coastal damage such as storm surge and gradual erosion. The '*ngaunga of Hine-moana* (the biting and gnawing of Hine-moana)'[17] results in a constant onslaught of sea into the land that eats away at the coastline. Hine-moana's unpreventable actions have created a volatile relationship between people and the coastline. The relationship must be constantly negotiated in order for people to continue to utilise the coastal regions. Coastal management requires strategies to prevent the resulting erosion and flooding, disruptions to road networks and communications, loss of private property and community assets, and the effects on water quality. One natural barrier to the sea is the sand dunes that fringe the coast and 'offer a buffer against the many storms that visit our shores'.[18] Many of these have undergone significant modification from real estate development and other human activities (such as pedestrian and vehicle access over fragile dune landscapes). The maintenance and restoration of the dunes has become an increasingly critical task'[19] that can be undertaken through council and community joint efforts and programmes.

One initiative is through Coast Care Bay of Plenty; The Bay of Plenty Regional Council in partnership with Tauranga City Council, Whakatane City Council, Western Bay of Plenty Council, Opotiki District Council and the Department of Conservation. The councils co-ordinate community action programmes that help to both reduce damage to sand dunes and assist with restoration. The community presence is through the 25 Coast Care Groups that include local residents, holiday homeowners, beachfront property homeowners, property developers, and conservationists. The Coast Care Groups monitor the dune health and advise the relevant council on work that they see as a priority. As with a lot of community initiatives in A/NZ, the restoration work is carried out on a volunteer basis. The Coast Care community groups carry out plantings, building of sand ladders for managed pedestrian access, and in some cases working with councils to remove or modify coastal storm water drainage to reduce erosion. The other important work is creating back-yard buffers, where beachfront homeowners are encouraged to 'restore native plant communities on dune areas adjacent to their properties'.[20]

The council-supported dune restoration programmes are an important strategy not only for vulnerable areas such as the Bay of Plenty coastline, but for all of New Zealand's coastal areas. In Dunedin's coastal suburb of Brighton, the DCC has initiated a pikao planting programme along the dune structures to ensure that the dunes remain an effective barrier against encroaching sea. Unlike the Coast Care example, the Brighton Beach restoration has its origins in understanding the dune system from an MEK perspective.

The story of the interspace between land and sea acknowledges the origin of pikao, a sedge grass found in coastal dune systems. The importance of pikao as a buffer between land and sea is recalled through a story of a battle between Tāne and Takaroa. The two fought for dominance over the environmental interface where land meets sea. With each surge from Takaroa, more of the land was eroded. Tāne fought back and eventually gave his eyelashes to Takaroa as a peace offering. Takaroa refused to accept them and threw them down where they grew to create an impenetrable barrier between sea and shore. The eyelashes became pikao and locked the sand into the coastline so creating the dunes and forming a barrier against the sea (Fig. 6.1).[21]

The DCC, in consultation with Otago-based Kāi Tahu, has erected notices around the area to provide information about the purpose of the pikao restoration. The notices serve to educate communities about the importance of this initiative in restoring the natural buffer zone between the community and the sea. They also serve to remind of the principle of maintaining strong relationships between the realms of ancestors and present generations. Dune reclamation relies on understanding the battle between the ancestors Tāne and Takaroa and how the increased incursion was overcome. The story reminds us of the need to respect and manage that relationship.

Fig. 6.1 Photo of DCC notification of pikao protection at Brighton Beach, 2015

CASE STUDY: INAKA RESTORATION IN THE WAIKOUAITI
RIVER CATCHMENT, OTAGO

In the East Otago region the local hapū Kāti Huirapa recognised the vulnerable state of one of their key mahika kai resources in the Waikōuaiti River. The river sits within the wider catchment area known as Matainaka, which is a place name that points to the importance of the rich resource found in the area—inaka (freshwater fish, whitebait).[22] The inaka have social, environmental, cultural, and economic importance to Kāti Huirapa people and the surrounding community. As a permanent reminder placed on the landscape, place names such as Matainaka are important indicators of environmental conditions and change over time. The name recalls the place; the place recalls the stories; the stories recall the whakapapa; and from that each group knows how they belong and fit within the environment. The stories also connect all the elements within the ecosystems and layer them in such a way to explain the interconnecting relationships.

Understanding the relationships allows for flexible decision-making and practices to ensure longevity. Although the name is now commonly restricted to one specific place, in the past it acted as a reference point for the entire resource catchment area surrounding two major rivers (Waikōuaiti and Waihemo) and the resulting wetlands that stretched 200 km along the coastline. Over time however, much of the wetlands were drained for agriculture and European settlements and now only comprise a small part of the overall area. The land-use modifications and destruction of the wetlands impacted on the important mahika kai and reduced habitat for the inaka. In 2016 Kāti Huirapa Rūnaka partnered with the New Zealand National Science Challenge, Bio Heritage, to develop a strategy for the future sustainability of their inaka in the Waikōuaiti river. Kāti Huirapa leads the project and the two researchers are from Kāti Huirapa. The starting point was the name, Matainaka, which indicates the purpose for mana whenua status over the catchment area and connects Kāti Huirapa people with their responsibilities and obligations to the longevity of the resource for future generations. Kāti Huirapa TEK acknowledges the part the river plays in the wider catchment area, and indicates how the whakapapa intertwine with intergenerational knowledge that has governed and managed the area in the past. Once the decision was made to find new management solutions for the inaka, a study was undertaken to identify the points at which the fishery would be at its most vulnerable. Two areas of the project that have been a focus for action are the potential increased flooding events and sea-level rise, which will increase the salt-water reach from the estuary. The NIWA climate change report allows Kāti Huirapa to work out how to gain most benefit through the future transitional states. Kāti Huirapa knows what will happen and they have an idea of when it may happen. The conditions for the relationship that maintains inaka habitat are heading for a transitional period that forces renegotiation to meet the changing environment. Any adjustments made will reflect the values, practices, and intergenerational focus to negotiate the most beneficial arrangement. In 2017 a major flooding event caused damage to the river and community, but did allow the Iwi to understand how future flooding events may impact on key spawning sites. This allowed for positive actions and planning that accounts for future water level rise, such as bank realignment, and corrective planting that best suits the spawning conditions. Observation of the local environment during the flooding event demonstrated an important lesson for the future activities. Overtime inaka have adjusted to the decreasing wetlands habitat by adopting new spawning strategies. They now favour the exotic grasses that line the banks in the upper reaches of the river. The exotic grasses provide a tight mesh of roots and foliage that helped the inaka survive the flood more successfully than the native grasses. The hapū had been replanting with native grasses, but will now opt to adjust the planting regime to meet the changed conditions initiated by the inaka. This has been an important lesson that teaches us to observe and understand how each part of an ecosystem responds to transitions in relationships-in this case the way that inaka renegotiated the relationship with the habitat. The project is ongoing and to date has succeeded in involving elders, younger tribal members and the wider community in taking a lead towards ensuring future social, cultural, environmental, and economic needs and aspirations are met.

These examples highlight some of the adjustments being undertaken in New Zealand and focus on both Iwi and New Zealand-wide approaches that benefit all communities. The three key adaptation vehicles are Iwi, central government, and LCs. All three have expectations, responsibilities, and obligations to find ways for future sustainability in the face of climate change. The participation of communities and individuals comes from the trickle down impact of local government policy, strategies, and planning. In line with two key climate change challenges (sea-level rise and flooding), one of the most common council-community-individual initiatives for coastal communities and cities is dune management. The Matainaka example demonstrates how the vulnerability of a key mahika kai resource was identified and future management strategies and planning will adjust to limit climate change impacts. The future adjustments are in line with Kāti Huirapa aspirations and expectations to ensure intergenerational development in line with cultural values and practices. MM recognises the interrelatedness between all elements with the ecosystem and how the relationships are managed as challenges occur. The transitions from one state to another in the ever-changing relationships stem from an understanding of how a particular ecosystem embraces change and the overall impact on the environment.

NOTES

1. McCarthy et al. 2010.
2. Kwadijk et al., 2010.
3. Kawadijk et al., 2010.
4. Nyong 2007.
5. Kwadijk et al., 2010.

6. King et al., 2012.
7. *New Zealand's Framework for Adapting to Climate Change*, Information Document from the Ministry for the Environment, www.mfe@govt.nz – Last accessed December 2014.
8. *New Zealand's Framework for Adapting to Climate Change*, Information Document from the Ministry for the Environment, www.mfe@govt.nz – Last accessed December 2014.
9. *New Zealand's Framework for Adapting to Climate Change*, Information Document from the Ministry for the Environment, www.mfe@govt.nz – Last accessed December 2014.
10. Ministry for the Environment, *Climate Change and Long-term Council Community Planning*, 2008, 2. [http://www.mfe.govt.nz/climatechange/]. Last retrieved 28 January 2015.
11. Ministry for the Environment *Climate Change and Long-term Council Community planning*, 2008, 2. [http://www.mfe.govt.nz/climatechange/]. Last retrieved 28 January 2015.
12. Ministry for the Environment, *Climate Change and Long-term Council Community Planning*, 2008, 3.
13. Ministry for the Environment, *Climate change and Long-term Council community Planning*, 2008, 9.
14. Dunedin City Council *Waste Management and Minimisation Plan*. [http://www.dunedin.govt.nz/whats-on/waste-management-and-mini-misation-plan/] Last retrieved 30 January 2015.
15. Dunedin City Council, *Waste management*, 2015, 3.
16. Mayor Dave Cull, commentary in *Otago Daily Times*, 13/06/15.
17. Brougham et al., 2009, 126.
18. Ministry for the Environment, *Community-based Dune Management for the Mitigation of Coastal Hazards and Climate Change Effects – A Guide for Local Authorities (2005)*, Case Study 1. [http://www.mfe.govt.nz/publications/climate/]. Last retrieved 28 January 2015.
19. Ministry for the Environment, *Community-based Dune Management... Case study 1.* [http://www.mfe.govt.nz/publications/climate/] Last retrieved 28 January 2015.
20. Bay of Plenty Regional Council, *What is Coast Care? Back Yard Buffers Project.* [http://boprc.govt.nz/].
21. Discussion among the elders at a Kati Huirapa meeting, October 2016. An art work by Ngāi Tahu artist Ewan Duff depicting the battle can be found in the Dunedin offices of the Department of Conservation.
22. Matainaka indicates two life-cycle stages of the *Galaxids* species of freshwater fish. *Mata* are the small fry which are harvested between September and October; *inaka* are the adult fish.

REFERENCES

Brougham, A. E., Reed, A. W., & Kāretu, T. (2009). *The Raupō Book of Māori Proverbs*. Auckland: Penguin Group (NZ).

King, D., Dalton, W., Duncan, M., Srinivasan, M. S., Bind, J., Zammit, C., McKerchar, A., Ashford-Hosking, D., & Skipper, A. (2012, March). *Maori Community Adaptation to Climate Variability and Change. Examining Risk, Vulnerability and Adaptive Strategies with Ngati Huirapa at Arowhenua Pa, Te Umu Kaha (Temuka), New Zealand.* Prepared for Te Runanga o Arowhenua Society Incorporated and the New Zealand Climate Change Research Institute – Victoria University, Wellington.

Kwadijk, J. C. J., Haasnoot, M., Mulder, J. P. M., Hoogvliet, M. M. C., Jeuken, A. B. M., van der Krogt, R. A. A., van Oostrom, N. G. C., Schelfhout, H. A., van Velzen, E. H., van Waveren, H., & de Wit, M. J. M. (2010). Using Adaptation Tipping Points to Prepare for Climate Change and Sea Level Rise: A Case Study in the Netherlands. In *WiREs Climate Change*. John Wiley & Sons Ltd. http://wiley.com/climatechange

McCarthy, J. J., Canziani, O. F., Leary, N. A., Dokken, D. J., & White, K. S. (Eds.). (2010). Cited in Kawadijk, J. C. J., Haasnoot, M., Mulder, J. P. M., Hoogvliet, M. M. C., Jeuken, A. B. M., van der Krogt, R. A. A., Oostrom, N. G. S., Schelfhout, H. A., van Velzen, E. H., van Waveren, H., & de Wit, M. J. M. Using Adaptation Tipping Points to Prepare for Climate Change and Sea Level Rise: A Case Study in the Netherlands. In *WIREs Clim Change*. Wiley.

Nyong, A., Adesina, F., & Osman Elasha, B. (2007). The Value of Indigenous Knowledge in Climate Change Mitigation and Adaptation Strategies in the African Sahel. *Mitigation and Adaptation Strategies for Global change, 12*, 787–797. https://doi.org/10.1007/s11027-007-9099-0.

Where to from Here? Learning from Our Pacific Neighbours

Abstract Awareness of climate change is increasing, as more is understood about change in Aotearoa/New Zealand (A/NZ) and across the Pacific. The Pacific strategies are the results of long-standing observation, planning, and action. Māori too utilise regionally based traditional knowledge frameworks to find solutions to new challenges and are well placed to lead attempts to lessen the impact from climate change in A/NZ. This chapter summarises the efforts so far, and also introduces new technological initiatives that could assist in the future such as blue carbon, salt-water resistant crops, and 'floating houses'. The Pacific Island Countries have shown the way and A/NZ can learn from the many ways that indigenous Pacific knowledge's have been utilised to help lessen the impact from climate change in the Pacific.

Keywords Blue carbon • Carbon capture and storage • Pacific lessons • Māori leadership • New technology initiatives

The climate change strategies utilised by New Zealand's Pacific neighbours are the results of long-standing observation, planning, and action. These are underpinned by TEK that utilises the locally referenced ecological knowledge with its history of adjusting to past environmental disaster to draw from. TEK is experiential and draws upon lessons from the past with the addition of intergenerational knowledge over time. The intergenerational

© The Author(s) 2019   85
L. Carter, *Indigenous Pacific Approaches to Climate Change*,
Palgrave Studies in Disaster Anthropology,
https://doi.org/10.1007/978-3-319-96439-3_7

aspect of IK is best summed up in the Kāi Tahu vision statement: 'Mō tātou, ā mō kā uri ā muri ake nei. For us and our children after us'.[1] All the planning, strategies, and knowledge that is utilised in contemporary situations is to ensure that the responsibilities and obligations to people and the environment can be carried sustainably into the future. A/NZ has its own set of tradition-based ecological knowledge—MEK—that references not only the environment but also the economic, social, and cultural variables as well. MEK places Māori in a leadership position to drive A/NZ forward in adapting to the challenges from climate change. MEK provides a platform from which to develop intergenerational strategies and practices that also draw on the experience of New Zealand's Pacific neighbours. MEK is based around how we form, manage, and negotiate relationships between people and their environment. Indeed we can learn from observing how the environment and the ecosystems within it (including human) adjust and transition to new states of compatibility. By considering that everything is monitored and managed through relationships formed over time, we can then revisit the climate change debate. There is a need to negotiate ways to repair relationships and work with the changing states that have been created. The challenge is to recognise and find ways to engage with the new environment and effect repairs necessary to have a compatible ecosystem. Reorganising the roles and responsibilities, and finding ways to maintain those, is an intergenerational role that requires flexibility in negotiating the ever-changing circumstances. The environment does not stand still and neither does our position in it remain static. MM, TEK, and IK all have processes to manage change and negotiate new states for sustainability.

PACIFIC RESPONSES

The adaptation measures discussed in Jenny Bryant-Tokalau's book demonstrate how TEK works in long held beliefs and practice as well as contemporary science. The Pacific Islands, using examples of land building, relocation, food storage, and preservation, as well as a deep traditional understanding of changing weather patterns and their impacts on sources of sustenance and the challenges of contemporary settlements are committed to a future of 'climate resilient green blue Pacific economies'.[2]

It is clear that the Pacific nations will become even more outspoken and vocal in their responses to the impacts of climate change. At the governance level, the Pacific is divided. Australia and New Zealand supported only limited action on climate change and faced strong verbal censure from some leaders, as well as from the Pacific Islands Development Forum (which specifically excludes both 'big brothers'). It has been claimed 'that a Pacific Islands Forum with Australia and New Zealand as members is hampering the ability of the Pacific island states to defend their interests, and in the case of climate change policy, their very survival'.[3,4]

Although the PIC are considered to be particularly vulnerable to climate change, what is also clear is that they have many ways of dealing with what now appears to be inevitable. Traditional practices, global lobbying, and new forms of adaptation will become not only more necessary, but also increasingly practiced. The responses are not merely knee-jerk reactions. Local scientists, working with communities will continue to search for local solutions, and at the regional level, new forms of governance are also emerging. The youthful Pacific Island Development Forum (PIDF) issued a strong declaration at the end of the Suva meetings. The Suva Declaration on Climate Change contained statements of concern and disappointment surrounding the impacts of climate change and the responses of the global community. The debate as to whether agricultural GHG emissions should be taxed through the NZETS system has long been debated. The problem has been part of the PIC protest about New Zealand's failure to take climate change seriously. Recent calls to have the issue addressed more effectively have again met with challenges from the industry. It is clear however that New Zealand's high contribution to GHG emissions through its agricultural industry have a severe impact on not only A/NZ, but more keenly on her Pacific neighbours. Some relatively new research concerns the development of low-carbon producing animal feed; and suggestions around the reduction in stock numbers to offset the increased GHG emissions from the agriculture sector—particularly dairy farming. These will not work in isolation, but rather alongside other measures that will drive mitigation and adaptation in this sector.

One form of mitigation that does exist within the NZETS is to increase carbon sinks to offset GHG. Carbon sinks with the capacity to increase CCS are an important part of mitigation strategies. In A/NZ this commonly occurs through reforestation allowances in the NZETS and is in fact the only CCS activity in the NZETS. The agriculture sector is currently only required to record the emissions, but not pay for them. Any CCS trade-offs in the farming sector through reforestation (among others such as improving degraded soils and wetlands) are not counted towards New Zealand's reduction in emissions. The loss of forest carbon sinks has been accepted into the mitigation policies around climate change, and feature prominently in carbon trading markets. In the NZETS there are inducements to replant forests that have been cleared for further development into agricultural land. Iwi have begun to investigate reforestation projects, but only if the knowledge frameworks within respective Mātauraka-a-Iwi can be the deciding development process. When the agriculture industry is included in the NZETS any contribution to improve CCS along with adaptation measures (adoption of low-carbon producing feeds and lower stock numbers), would go someway to making the ETS an effective mitigation measure. There is another 'hidden' source for CCS that exits in agricultural lands as well as Iwi-held coastal environments and these are collectively known as blue carbon areas.

Increasing CCS

Blue carbon sinks exist in the biomass and sediments of habitats such as mangroves, sea grass beds, tidal marshes, and other marine and coastal vegetated ecosystems'.[5] When blue carbon areas are degraded or destroyed, they release the carbon into the atmosphere. Wetland drainage for increased agricultural land has been ongoing in A/NZ since the nineteenth century and wetlands have been 'reduced to about 10% of their original extent'.[6] Marjo Vierros noted that, as a new research area, it is unknown as to how much carbon is stored in coastal and marine vegetation systems, but it is known that 'coastal ecosystems are among the most intense carbon sinks on the planet'.[7] The blue carbon areas have not been included into carbon trading schemes, despite the coastal ecosystems contributing significantly to atmospheric GHGs, and potential significant economic and environmental benefits from better management.[8] In times when carbon trading is becoming more common, there is huge potential economic benefit for indigenous and local communities whose territories include the estuarine and coastal wetlands environments. The potential environmental benefits in protecting and sustainably managing the estuarine and coastal ecosystems also ultimately contribute to the economic benefits. For example, the seafood market value of mangroves in the United States coastal areas has been reported at US$7,500–$167,500 per km².[9] In the past indigenous peoples fully utilised these areas and many of the resources were a big part of the economic and social life for Māori and other Pacific peoples. There is potential for coastal estuaries, wetlands, and sea grass meadows to replace current food development areas by focusing on alternative food sources held within them. This does not deny the complexities of each wetland area on a global scale. Each will have its own changes in hydrology, increased temperatures, flooding, and ecosystem changes.[10] They are region specific and understood best by the people who live among and alongside them and as Erwin argues each 'needs to be viewed in the broader context of their spatial location in a watershed within a specific region'.[11]

There is a commonly held belief that degraded wetlands must be restored as part of the solution to reduced carbon storage areas. While preventing further degradation is a good start, Irving et al. have suggested restoration is not the key to increased CCS. In all cases the degraded areas have already lost their CO_2 into the atmosphere and the costs to restore and build up the CCS capability of existing degraded systems is prohibitive. Rather, preventing further habitat loss and encouraging natural recovery is a more cost-efficient solution.[12] Protection measures for remaining wetlands would benefit not only the ecosystems, but also act to prevent accelerated discharge of the carbon deposits into the ocean and atmosphere. A/NZ is in a position to lead the way for the inclusion of blue carbon sinks into ETSs. This would take the emphasis off solely economic development and encourage more intergenerational and sustainable land-use development practices. It would also encourage innovation in new resource development that echoes the past practices and food types common to all coastal Iwi.

The relatively new approaches around blue carbon science are echoed in the recent scientific studies on the value of organic carbon burial in fjord sediments.[13] In their research paper that details carbon sequestration in fjords, Smith et al. discuss the potential for 'fjords to play an important role in climate regulation on glacial-interglacial timescales'.[14] The long-term burial of organic carbon in marine sediments has played a key role in controlling atmospheric O_2 and CO_2 concentrations and the fjords 'have been hypothesized to be hotspots of organic carbon burial, because they receive high rates of organic material fluxes from the watershed'.[15] New Zealand has a large fjord system in Fiordland National Park. The Park makes up part of the south-west region of Te Tiritiri o te Moana (the Southern Alps) in Te Wai Pounamu/South Island and contains 14 fjords. The fjords act as a conduit for sediments from 'mass-wasting events (landslides) caused by steep slopes, shallow soils, and seismic activity'[16] that erode organic carbon into the ocean. Despite numerous studies that have demonstrated the delivery of organic carbon into the oceans via fjords, they have 'largely been ignored as a major global depositional environment'.[17] The slow-release of organic carbon, like its blue carbon counterparts, is part of the natural process for GHG into the atmosphere. The anthropogenic-caused acceleration of this will speed up the process and add to the impact from climate change in areas such as sea-level rise, ocean acidification, and ultimately contribute to the impact on sea and marine-based economic activities. The new research for ways to increase and/or stabilise CCS in the coastal and marine environments is future proofing strategies and practices to minimise the impact from climate change. New Zealand with its proliferation of blue carbon areas and fjords could recognise the potential of these areas as long-term carbon sequestration areas, along with existing land-based reforestation policies. An Open Statement to COP16 on Blue Carbon solutions to Climate Change in 2010,[18] called for the establishment of a Blue Carbon Fund. It would be wise for A/NZ to revisit, take notice of and emulate the CCS attempts in the Pacific. Two examples from Sovi in Fiji and Vanuatu stand out as potential instructive cases for A/NZ in combining both mitigation and adaptation as hand-in-hand activities and strategies.[19]

Pacific Blue Carbon: Case Studies

For the past three decades there has been interest shown in the future of the Sovi Basin, a forested and diverse ecosystem of around 19,600 hectares (50,000 acres) on the large island of Viti Levu. This quite remote area has been considered at various times as an important carbon sink, and most recently as an area worth protecting under the UNESCO World Heritage cultural category. In a partnership between Conservation International and Fiji Water, a bottled water company based in Fiji, there is an initiative to conserve and protect the area as one of the few remaining virgin tropical forests in inland Fiji. In response to Fiji's commitments to the UNFCCC, the Sovi Basin carbon sequestration programme aimed to promote sustainable management, and promote and cooperate in the conservation and enhancement of sinks and reservoirs of all GHGs. There was also the addition of the Sovi Basin Trust Fund (SBTF), which is supported by donations from Conservation International. This fund makes annual disbursements that not only offset the cash value of logging pay outs to local land owners, but also pay for land leases and create jobs. Such an approach is considered to be innovative and unusual in the Fiji context, but issues of land tenure continue to raise problems as landowners of the resource and the partners cannot come to an agreement over lease arrangements.[20] The fact that Sovi Basin remains on the tentative list for UNESCO, and has not become a successful example of carbon sink sequestration is a pity, but does point to the complexity and time required for PIC in resolving issues surrounding land tenure.

More recent plans for Vanuatu to realise the potential of carbon sinks[21] are underway following a scoping study by the Commonwealth Secretariat. This work aims to showcase what is possible in blue carbon, but cautions that it is not only to do with financial reward, but instead looks towards community and ecosystem benefits, crucial for livelihoods in smaller countries, particularly those with large areas of mangrove. Vanuatu had examined issues of carbon sinks before and organisations, including governments, were aware of the risky nature of carbon trading, but in the past two years, working closely with other projects, such as REDD+ is recruiting staff and establishing mangrove projects in what was well become a positive example in the Pacific towards managing coastal (blue) carbon sinks.

What the energy behind such projects demonstrate, is that time, scepticism and failure to act are not options for the smaller PIC. Given that the emissions of Pacific countries are negligible, the human energy that has already gone into the Sovi and Vanuatu projects may still act as a lesson to A/NZ.

FUTURE INNOVATIONS: TEK AND TECHNOLOGY

The IPCC reports recommended TEK in that are future focused and may involve solutions that have not as yet been seriously considered. In Chap. 2 the discussion around TEK/MEK included the development and sustainability of relationships in the past, present, and future generations. This will involve establishing relationships with groups who can provide technical innovations for managing vulnerable resources and landscapes. In MEK the principle of whanaukataka ensures all relationships have longevity and are mutually beneficial across changing circumstances and through intergenerational development. Some of the desired relationships are with non-indigenous peoples and organisations in a knowledge sharing capacity for economic, social, environmental, and cultural development. There are a number of technology-based adaptation solutions being developed across the world, many of which could benefit PIC including A/NZ. These future focused adaptation innovations such as salt-water tolerant crops, and 'floating' houses could prove to be valuable additions to the A/NZ suite of strategies.

Future food security is a topical issue in times of climate change for PIC and A/NZ. This is not a new challenge in the Pacific areas where countries have had to develop innovative solutions to changing environmental conditions over time. The new challenge coming from sea-level rise will impact on salinity levels in freshwater reserves thus impacting on irrigation for arable lands. The United Nations Food and Agriculture Organisation has reported that salination is reducing the world's irrigated lands by 1–2 percent annually. In response, scientists in the Netherlands are experimenting with using saline water for irrigation and so far have proven that some crops will grow successfully in saline water. Van Rijsselberghe owner of the test farm worked with scientists from the Free University of Amsterdam, and has managed to grow 'carrots, cabbage, onions and beetroot, but potatoes proved to be the most tolerant to saline water'.[22] The PIC are experiencing increased salinity in the water tables and experiments such as this could prove to be invaluable for future inhabitants. A/NZ could also benefit from the knowledge gained, especially coastal Iwi and other communities who face similar increased salinity problems. Food security is one area where adaptation needs to be innovative and future focused to meet the challenges from climate change. In the past the PIC and Māori tribal groups have developed ways to adapt to living in areas that succumb to flooding and sea water inundation. These include land creation and building houses on stilts to combat increased flooding.[23] Communities in Fiji and Vanuatu have also introduced small-scale sea walls to protect housing and land; and planted grasses and trees along the coastline. In A/NZ, sea walls are being investigated as part of any future adaptation measures. Although PIC have used a combination of housing relocations along with sea walls to stop erosion, there is one initiative that could offer the technology for Pacific adaptation to sea-level rise and increased flooding.

As an adaptation measure against increased flooding and rising sea levels the technology to develop floating houses is currently being tested in places such as The Netherlands, Florida (post Hurricane Katrina) and London (Thames riverside site). The London house designed by Baca Architects is constructed to float and rise along with the water levels by raising and lowering on four guideposts within a fixed dock.[24] Another example of floating homes technology is in Amsterdam, the Netherlands. The new suburb of Ijburg has 36 houses built on Steigreiland, one of several man-made islands in an artificial lake. The base of the house is filled with cement and heavy-duty foam. Rings attached to sunken posts make the houses stay put as it rises up and down, up to seven and half meters above the water level.[25] The Netherlands 'faces sea-level rise by up to 1.3

meters in the next century and up to 4 meters over the next 200 years'.[26] One third of the Netherlands is at sea level or below it, and as Pavel Kabat of the Dutch Government's Delta Commission explains, 'you can't just solve the problem with dykes, we have to change strategy...we shouldn't see water as a danger, but as a chance, as a challenge'.[27] The futuristic building innovations being utilised in the Netherlands (and London) provide worthy technological options for coastal community inundation here in A/NZ. Advances in technology form part of a mix of adaptation strategy's needed to put A/NZ on the pathway to living with the impact from climate change, particularly in A/NZ vulnerable coastal regions. A solution that works with the changing environment and climate is one way that MEK-based solutions can be realised here in A/NZ. The atua in charge of land, sea, and weather are forever in an uneasy truce; and as such, people must continuously find ways to adapt and live within their respective realms and environments.

PIC have demonstrated through their many efforts at adaptation and sheer determination, that there are ways of coping that involve significant cross party agreements, major legislation, and virtually no scepticism. The impacts of climate change, exacerbated by the behaviour of humans throughout the world, are now affecting not only the low-lying atolls of the Pacific, but A/NZ. The issue is not one for PIC alone, but requires their larger industrialised neighbour, A/NZ, to step up and take responsibility for developing and actioning effective mitigation and adaptation policies. Indigenous peoples forging relationships with climate change scientists and environmental scientists goes some way to meeting the IPCC recommendation for the utilisation of indigenous knowledge in climate change adaptation. The relationships allow a uniquely local voice that carries generations of experiential knowledge and practices based around close connections with the environment. As we complete these books, we acknowledge the release of the October 2015 New Zealand State of the Environment Report[28]; the very recent calls for the New Zealand agriculture industry to take responsibility for the industry's GHG emissions (October 2015); and more importantly the change of national Government (2017). The incoming climate change minister, James Shaw, has indicated that adaption will now become a focus to work alongside mitigation. These changes within A/NZ stand as evidence that the climate change debate is evolving and developing as more evidence for the rise of anthropogenic climate changes takes hold. The two books reach no final conclusions in the climate change debate—rather we suggest that they open the conversation. The books emphasise the ongoing challenge of climate change and potential for effective mitigation and adaptation as demonstrated through TEK. The basis of TEK ensures innovative, intergenerational policies and practices arise from past and present experiences and solutions. PIC have shown the way for the two key identified climate change threats: sea-level rise and flooding. They are also thinking forward by investing time and money into innovative new ways of increasing CCS areas—again lessons that A/NZ could learn from. Perhaps though it is time that A/NZ realised that climate change is not everyone else's problem and took heed of the long-term resilience exhibited by indigenous peoples. A/NZ cannot afford to ignore the wealth of intergenerational knowledge and practices that have accumulated intergenerationally across the Pacific, and by utilising MEK traditional strengths and knowledge, A/NZ 'can face this new assault of climate change head-on'.[29]

NOTES

1. Ngāi Tahu Iwi mission statement, in Development strategy *Vision 2015*. 'www.ngaitahu.iwi.nz'.
2. The Third Pacific Islands Development Forum Leaders Summit, held in Suva in September 2015 had as its overarching theme "Building Climate Resilient Green Blue Pacific Economies".
3. G. Fry, http://devpolicy.org/pacific-climate-diplomacy-and-the-future-relevance-of-the-pacific-islands-forum-20150904/
4. It is fully understood that despite the absence of Australia and New Zealand at the PIDF meetings, other powers, such as China, were welcomed.
5. Vierros, 2013.
6. Ministry for the Environment, 'New Zealand's Environment at a glance', *Environment Aotearoa 2015*, 4 (http://www.mfe.govt.nz/publications/environmental-reporting/environment-aotearoa-2015/] Last accessed 27/10/2015.
7. Vierros, 2013.
8. Vierros, 2013.
9. Vierros, 2013.
10. Erwin, 2009, 72.
11. Erwin, 2009, 72.
12. Irving et al., 2001, 1.
13. Smith et al., 2015, 1–5.
14. Smith et al., 2015, 1.
15. Smith et al., 2015, 1.
16. Smith et al., 2015, 1.

17. Smith et al., 2015, 1.
18. Blue Climate Coalition, 'Blue Carbon Solutions for Climate Change' Open statement to the delegates of COP16. *blueclimatesolutions.org* November 30, 2010.
19. The two Pacific examples come with the courtesy of Jenny Bryant-Tokalau.
20. Final Report National Assessment, Climate Change United Nations Framework Convention On Climate Change (UNFCCC), Department of Environment, NCSA Project Steering Committee, NCSA Project and UNDP Fiji, October 2008.
21. D. d'A. Laffoley, 'The management of coastal carbon sinks in Vanuatu: realising the potential'. A report to the Government of Vanuatu. Commonwealth Secretariat, London. Commonwealth Secretariat, London, 2013.
22. Putic Greg, 'Dutch Experiment shows farming with salty water possible' in *Silicon Valley Technology*, http://www.voanews.com/a/farming-with-salty-water-is-possible/..., accessed 13/02/2017.
23. Bryant-Tokalau, 2018; McNamara and Prasaa, 2015.
24. 'UK's first amphibious house', in *Inhabitat.* http://inhabitat.com/6-amphibious-houses-that-float-to-escape-flooding/. Accessed 27/02/2017.
25. Kabat, P. 'Floating houses to fight climate change in Holland', in http://www.dw.com/en/floating-houses-to-fight-climate-change-in-holland/. Accessed 27/02/2017.
26. Kabat, P. 'Floating houses to fight climate change in Holland', in http://www.dw.com/en/floating-houses-to-fight-climate-change-in-holland/. Accessed 27/02/2017.
27. Kabat, P. 'Floating houses to fight climate change in Holland', in http://www.dw.com/en/floating-houses-to-fight-climate-change-in-holland/. Accessed 27/02/2017.
28. Ministry for the Environment, *Environment Aotearoa 2015* [http://www.mfe.govt.nz/publications/environmental-reporting/environment-aotearoa-2015].
29. Parker et al., 2006, 29.

REFERENCES

Bryant-Tokalau, J. (2018). *Indigenous Pacific Approaches to Climate Change: Pacific Island Countries.* Palgrave Studies in Disaster Anthropology Series: Palgrave Pivot. Cham: Springer International Publishing. e-book: https://doi.org/10.1007/978-3-319-78399-4.

Erwin, K. L. (2009). Wetlands and Global Climate Change: The Role of Wetland Restoration in a Changing World. *Wetlands Eco Management, 17,* 71–84.

Irving, A., Connell, S. D., & Russell, S. D. (2011, March 29). Restoring Coastal Plants to Improve Global Carbon Storage: Reaping What We Sow. *PLoS One, 6*(3), e18311. https://doi.org/10.1371/journal.pone.0018311.

McNamara, K. E., & Prasaa, S. S. (2015). Valuing Indigenous Knowledge for Climate Change Adaptation Planning in Fiji and Vanuatu. In *Traditional Knowledge Bulletin, Tropical Issues Series.* Traditional Knowledge Initiative of the United Nations University – Institute of Advanced Studies. http://www.unutki.org/

Parker, A., Grossman, Z., Whitesell, E., Stephenson, B., Williams, T., Hardison, P., Ballew, L., Burnham, B., & Klosterman, R. (Eds.). (2006). *Climate Change and Pacific Rim Indigenous Nations.* Washington, DC: Northwest Indian Applied Research Institute (NIARI), The Evergreen State College, Olympia.

Smith, R. W., Bianchi, T. S., Allison, M., Savage, C., & Galy, V. (2015). High Rates of Organic Carbon Burial in Fjord Sediments Globally. *Nature Geoscience.* Advance on-line publication, www.nature.com/naturegeoscience

Vierros, M. (2013, October 10). Communities and Blue Carbon: The Role of Traditional Management Systems in Providing Benefits for Carbon Storage, Biodiversity Conservation and Livelihoods. *Springer Science+Business Media Dordrecht, Special Edition: Climate Change.* https://doi.org/10.1007/s10584-013-0920-3. Last accessed 2015.

References

Arctic Council. (2005). *Arctic Climate Assessment Report*. New York: Cambridge University Press.

Barnett, J., & Campbell, J. (2010). *Climate Change and Small Island States: Power, Knowledge and the South Pacific*. London: Earthscan.

Beckwith, M. (1970). *Hawaiian Mythology*. Honolulu: University of Hawaii Press.

Berkes, F. (2012). *Sacred Ecology* (3rd ed.). New York: Routledge.

Boillat, S., Serrano, E., Rist, S., & Berkes, F. (2012). The Importance of Place Names in the Search for the Ecosystem-Like Concepts in Indigenous Societies: An Example from the Bolivian Andes. *Environmental Management* (2013), *51*, 663–678 (on-line version). https://doi.org/10.1007/s00267-012-9969-4.

Brougham, A. E., Reed, A. W., & Kāretu, T. (2009). *The Raupō Book of Māori Proverbs*. Auckland: Penguin Group (NZ).

Bryant-Tokalau, J. (2014, December 3). Indigenous Responses to Environmental Challenges: Artificial Islands and the Challenges of Relocation. Paper presented session, 'Climate Change, Disasters and Pacific Agency' *Pacific History Conference, Lalan Chalan Tala Ara*, Taipei, Taiwan.

Bryant-Tokalau, J. (2016). *Human & Environmental Security: What the Pacific Can Teach NZ & Australia About Climate Change*. Macmillan Brown Center for Pacific Studies: Pacific Policy Brief 2016/2 ISSN 1172-3416.

Bryant-Tokalau, J. (2018). *Indigenous Pacific Approaches to Climate Change: Pacific Island Countries*. Palgrave Studies in Disaster Anthropology Series: Palgrave Pivot. Cham: Springer International Publishing. e-book: https://doi.org/10.1007/978-3-319-78399-4.

© The Author(s) 2019
L. Carter, *Indigenous Pacific Approaches to Climate Change*,
Palgrave Studies in Disaster Anthropology,
https://doi.org/10.1007/978-3-319-96439-3

Buck, P., (Te Rangihiroa). (1954 [1975 reprint]). *Vikings of the Sunrise*. Christchurch: Whitcomb and Tombs Ltd.

Bullock, D. (2009). *The New Zealand Emissions Trading Scheme: A Step in the Right Direction?* (Institute of Policy Studies Working Paper 09/04, March 2009). Wellington: School of Government Studies, University of Victoria.

Cajete, G. (2000). *Native Science Natural Laws of Interdependence*. Santa Fe: Clear Light Publishers.

Carter, L. (2003). The Bureaucratisation of Genealogy. In *Ethnologies Compares* (Vol. 6). Paris: University of Montpellier. http://alor.univ-montp3.fr/cerce/r6/1.w.htm. Last accessed 2 July 2003.

Carter, L. (2004a). Naming to Own. Place Names as Indicators of Human Interaction with the Environment. In *AlterNative. An International Journal of Indigenous Scholarship*, issue *1*, 7–25. Auckland: Nga Pae o Te Maramatanga/The National Institute of Research Excellence in Māori Development, University of Auckland.

Carter, L. (2004b). *Whakapapa and the State. Some Case Studies in the Impact of Central Government on Traditionally Organised Māori Groups* (Unpublished PhD Thesis). Auckland: University of Auckland.

Carter, L. (2014, December 4–7). *We Are Not Drowning, Pacific Identity and Cultural Sustainability in the Era of Climate Change*. Conference presentation, Pan Pacific Indigenous Resource Management, Panel 3 – 'Managing the Sea and Land in Times of Climate Change', at Pacific History Association Conference, Taipei and Taitung, Taiwan.

Carter, L., & Ruru, J. (2005). "Freeing the Natives" The Role of the Treaty of Waitangi in the Reassertion of tikanga Māori. In N. Thomas (Ed.), *Te Tai Haruru. Journal of Maori Legal Writing*, *2*, 15–36.

Carter, L., Kamou, R., & Barrett, M. (2011). *Literature Review and Programme Report. Te Pae Tawhiti Maori Economic Development Porgramme*. Published Report for Nga Pae o Te Maramatanga, University of Auckland.

Chand, S., Chambers, L., Waiwai, M., Malsale, P., & Thompson, E. (2014). Indigenous Knowledge for Environmental Prediction in the Pacific Island Countries. *American Meteorological Society*. https://doi.org/10.1175/WCAS-D-13-00053.1.

Clarke, W. C. (1990). Learning from the Past: Traditional Knowledge and Sustainable Development. *The Contemporary Pacific*, *2*(2), 233–253.

Cowan, J. (1923). The Story of Niue. Genesis of a South Sea Island. *The Journal of the Polynesian Society*, *32*(128), 238–243.

Craig, R. D. (1989). *Dictionary of Polynesian Mythology*. New York: Greenwood Press.

Crooks, S., Herr, D., Tamelander, J., Laffoley, D., & Vandever, J. (2011). *Mitigating Climate Change Through Restoration and Management of Coastal Wetlands and Near-Shore Marine Ecosystems. Challenges and Opportunities* (Environment Department Paper 121). Washington, DC: World Bank.

Davidson, J. (1992). *The Prehistory of New Zealand*. Auckland: Longman Paul Limited.

Davidson-Hunt, J., Julian Idrobo, C., Pengelly, R. D., & Sylvester, O. (2013). Anishinaabe Adaptation to Environmental Change in Northwestern Ontario: A Case Study in Knowledge Coproduction for Non-timber Forest Products. *Ecology and Science, 18*(4), 44. http://www.ecologyandsociety.org/vol18/iss4/art44.

Doherty, W. (2010). *Mātauranga a Tūhoe. The Centrality of Mātauranga-a-iwi to Māori Education.* Unpublished PhD Thesis. University of Auckland. http://hdl-handle.net/2292/5639.

Egeru, A. (2012). Role of Indigenous Knowledge in Climate Change Adaptation: A Case Study of the Teso Sub-region, Eastern Uganda. *Indian Journal of Traditional Knowledge, 11*(2), 217–224.

Erwin, K. L. (2009). Wetlands and Global Climate Change: The Role of Wetland Restoration in a Changing World. *Wetlands Eco Management, 17*, 71–84.

Fuary, M. (2009). Reading and Riding the Waves: The Sea as Known Universe in the Torres Strait. *Historic Environment, 22*(1), 32–37.

Harmsworth, G. (2003). *Maori Perspectives on Kyoto Policy: Interim Results. Reducing Greenhouse Gas Emissions from the Terrestrial Biosphere (C09X0212).* Discussion Paper for Policy Agencies (Climate Change Office; MfE, MAF, TPK). Palmerston North: Landcare Research NZ.

Helander, E. (1999). Sami Subsistence Activities – Spatial Aspects and Structuration. *Acta Borealia, 16*(2), 7–25. https://doi.org/10.1080/0800389908580495.

Hopkins, D., Campbell-Hunt, C., Carter, L., Higham, J., & Rosin, C. (2015). Climate Change and Aotearoa New Zealand. *WIRes Climate Change.* https://doi.org/10.1002/wcc.355.

Hutchings, J. (2006). Negotiating (Bi)cultural Environmental Management Under the Resource Management Act. In M. Mulholland (Ed.), *State of the Māori Nation: Twenty-First-Century Issues in Aotearoa* (pp. 95–105). Auckland: Raupo Publishers.

Insley, C., & Meade, R. (2008). *Māori Impacts from the Emissions Trading Scheme. Detailed Analysis and Conclusions.* Wellington: Ministry for the Environment. Prepared by 37 Degrees South and The Cognitas Advisory Services.

Irving, A., Connell, S. D., & Russell, S. D. (2011). Restoring Coastal Plants to Improve Global Carbon Storage: Reaping What We Sow. *PloS ONE, 6*(3), e18311. https://doi.org/10.1371/journal.pone.0018311.

Kawharu, I. H. (Ed.). (1989). *Waitangi. Maori and Pakeha Perspectives of the Treaty of Waitangi.* Oxford: Oxford University Press.

Kawharu, M. (Ed.). (2002). *Whenua. Managing Our Resources.* Auckland: Reed Books Ltd.

King, D. N. T., Skipper, A., & Tawhai, W. B. (2007). Māori Environmental Knowledge of Local Weather and Climate Change in Aotearoa-New Zealand. *Climate Change.* Springer Science + Business Media BV, 385–409. https://doi.org/10.10077/s10584-007-9372-y.

King, D., Dalton, W., Duncan, M., Srinivasan, M. S., Bind, J., Zammit, C., McKerchar, A., Ashford-Hosking, D., & Skipper, A. (2012, March). *Maori Community Adaptation to Climate Variability and Change. Examining Risk, Vulnerability and Adaptive Strategies with Ngati Huirapa at Arowhenua Pa, Te Umu Kaha (Temuka), New Zealand.* Prepared for Te Runanga o Arowhenua Society Incorporated and the New Zealand Climate Change Research Institute – Victoria University, Wellington.

Klein, R. J. T., Schipper, E. L., & Dessai, S. (2003). Integrating Mitigation and Adaptation into Climate Change Development Policy. In N. Stehr & H. von Storch (Eds.), *Environmental Science and Policy, 8*(6), 579–588, December 2005.

Kwadijk, J. C. J., Haasnoot, M., Mulder, J. P. M., Hoogvliet, M. M. C., Jeuken, A. B. M., van der Krogt, R. A. A., van Oostrom, N. G. C., Schelfhout, H. A., van Velzen, E. H., van Waveren, H., & de Wit, M. J. M. (2010). Using Adaptation Tipping Points to Prepare for Climate Change and Sea Level Rise: A Case Study in the Netherlands. In *WiREs Climate Change.* John Wiley & Sons Ltd. http://wiley.com/climatechange.

Local Government New Zealand. (2004). *Local Authority Engagement with Māori: Survey of Current Council Practices July 2004.* Wellington: Local Government New Zealand.

Matunga, H. (2000). Decolonising Planning: The Treaty of Waitangi, the Environment and a Dual Planning Tradition. In A. Memon & H. Perkins (Eds.), *Environmental Planning and Management in New Zealand.* Auckland: Dunmore Press.

McCarthy, J. J., Canziani, O. F., Leary, N. A., Dokken, D. J., & White, K. S. (Eds.). (2010). Cited in Kawadijk, J. C. J., Haasnoot, M., Mulder, J. P. M., Hoogvliet, M. M. C., Jeuken, A. B. M., van der Krogt, R. A. A., Oostrom, N. G. S., Schelfhout, H. A., van Velzen, E. H., van Waveren, H., & de Wit, M. J. M. Using Adaptation Tipping Points to Prepare for Climate Change and Sea Level Rise: A Case Study in the Netherlands. In *WIREs Clim Change.* Wiley.

McFadgen, B. (2007). *Hostile Shores. Catastrophic Events in Prehistoric New Zealand and Their Impact on Maori Coastal Communities.* Auckland: Auckland University Press.

McNamara, K. E., & Prasaa, S. S. (2015). Valuing Indigenous Knowledge for Climate Change Adaptation Planning in Fiji and Vanuatu. In *Traditional Knowledge Bulletin, Tropical Issues Series.* Traditional Knowledge Initiative of the United Nations University – Institute of Advanced Studies. http://www.unutki.org/

McNamara, E. K., & Westoby, R. (2011). Local Knowledge and Climate Change Adaptation on Erub Island, Tores Strait. *Local Environment, 16*(9), 887.

Mead, H. M., & Grove, N. (2001). *Ngā Pēpeha a ngā Tīpuna. The Sayings of the Ancestors.* Wellington: Victoria University Press.

Mihinui, B. (2002). Hutia to rito o te harakeke. A Flaxroot Understanding of Resource Management. In M. Kawharu (Ed.), *Whenua. Managing Our Resources* (pp. 21–33). Auckland: Reed Books.

Ministry for the Environment. (2000). *Reporting on the Key Issues and Good Practice Guidelines for Local Authorities to Better Understand Statutory Responsibilities*. Wellington: Ministry for the Environment.

Nunn, P. D. (2003). Fished Up or Thrown Down: The Geography of Pacific Island Origin Myths. *Annuls of the Association of American Geographers, 93*(2), 350–364.

Nyong, A., Adesina, F., & Osman Elasha, B. (2007). The Value of Indigenous Knowledge in Climate Change Mitigation and Adaptation Strategies in the African Sahel. *Mitigation and Adaptation Strategies for Global Change, 12*, 787–797. https://doi.org/10.1007/s11027-007-9099-0.

Orange, C. (1987). *The Treaty of Waitangi*. Wellington: Allen and Unwen/Port Nicholson Press.

Park, G. (1995). *Ngā Uruora (The Groves of Life). Ecology and History in a New Zealand Landscape*. Wellington: Victoria University Press.

Parker, A., Grossman, Z., Whitesell, E., Stephenson, B., Williams, T., Hardison, P., Ballew, L., Burnham, B., & Klosterman, R. (Eds.). (2006). *Climate Change and Pacific Rim Indigenous Nations*. Washington, DC: Northwest Indian Applies Research Institute (NIARI), The Evergreen State College, Olympia.

Posey, D. (2002). Upsetting the Sacred Balance. Can the Study of Indigenous Knowledge Reflect Cosmic Connections? In P. Sillitoe, A. Bicker, & J. Pottier (Eds.), *Participating in Development. Approaches to Indigenous Knowledge* (p. 39). London: Routledge.

Ruckstuhl, K., Gale, K., Carter, L., Ellisson, E., Flack, S., & Russell, K. (2017). *Kā Rūnaka Expectations for Oil and Gas Companies in East Otago*. Dunedin: Kai Tahi Ki Otago Ltd.

Ruru, J. (2013). Te Ture – Mineral Law and Māori. In K. Ruchstuhl, L. Carter, L. Easterbrook, A. R. Gorman, H. Rae, J. Ruru, D. Ruwhiu, J. Stephenson, A. Suszko, M. Thompson-Fawcett, & R. Turner (Eds.), *Māori and Mining*. University of Otago: Māori and Mining Research Team. http://otago.ourarchive.ac.nz.

Sillitoe, P. (1998, April). The Development of Indigenous Knowledge: A New Applied Anthroplogy. *Current Anthropology, 39*(2), 223–252.

Smith, R. W., Bianchi, T. S., Allison, M., Savage, C., & Galy, V. (2015). High Rates of Organic Carbon Burial in Fjord Sediments Globally. *Nature Geoscience*. Advance on-line publication, www.nature.com/naturegeoscience.

The New Zealand Herald. (2014). *NZ's Emissions Impossible*. http://www.nzherald.co.nz/news/print.cfm?pbjectid=11374647. Last retrieved January 15, 2015.

Thornton, A. (1992). *Maui tiketike-a-taranga*: Te Rangikaheke to Grey. *GNZMMSS, 43*, 896–973; Auckland Public Library Collection. This version is also reprinted in Agathe Thornton, *The Story of Maui by Te Rangikaheke*. Edited with translation and commentary by A. Thornton. Christchurch: University of Canterbury Māori Studies.

Tol, R. S. J. (2005). Adaptation and Mitigation: Trade-Offs in Substance and Methods. In N. Stehr & H. von Starch (Eds.), *Environmental Science and Policy, 8*(6), 572–578.

Tranter, B., & Booth, K. (2015, July). Scepticism in a Changing Climate: A Cross-National Study. *Global Environmental Change, 33,* 154–164. http://www.sciencedirect.com/science/article/pii/S095937801500758. Accessed 15 July 2015.

Vierros, M. (2013, October 10). Communities and Blue Carbon: The Role of Traditional Management Systems in Providing Benefits for Carbon Storage, Biodiversity Conservation and Livelihoods. *Springer Science+Business Media Dordrecht, Special Edition: Climate Change.* https://doi.org/10.1007/s10584-013-0920-3. Last accessed 2015.

Wilbanks, T. J. (2005). Issues in Developing a Capacity for Integrated Analysis of Mitigation and Adaptation. In N. Stehr & H. von Storch (Eds.), *Environmental Science and Policy, 8*(6), 541–547.

Index[1]

A

Aotearoa/New Zealand (A/NZ),
v–vii, 1–4, 6–10, 15–18, 20, 21,
26, 27, 29, 32–35, 39–50, 55–67,
71–82, 86–94
  Karitane, 49
  North Island, 6, 31, 32, 39–41, 49,
65, 66
  South Island, v, vii, 6, 27, 32,
37n38, 39, 42, 50, 89
  Te Ika a Māui, 32, 41
  Te Wai Pounamu, 32, 89
  Waikōuaiti river, 49, 80–82
Arctic Climate Assessment Report
(ACCR), 4, 5
Atua, vi, 20, 29–35, 93
Atua, land
  Aoraki, 32
  Kane, 30
  Kupe, vi, 19
  Māui, 7, 31, 32
  Papatūanuku, 29, 32, 33

  pikao, 79, 80
  Tāne, 29–31, 33, 79
  Ti'iti'i, 31
  Tūterakiwhanoa, 32
  Wuhngin, 31
Atua, sea
  Kanaloa, 30
  Takaroa, 29–33, 79
  Tangaroa, 30, 33
Atua, weather
  Leomatagi, 33
  Ra'a, 33
  Tapakaumatagi, 33
  Tāwhirimatea, 7, 33
  Vinaka, 33

B

Berkes, F., 6, 26, 27, 34
Booth, K., 2
Bryant-Tokalau, J., v, vii, 3, 4, 6, 86
Bullock, D., 9, 56, 58, 66

[1] Note: Page numbers followed by 'n' refer to notes.

C

Climate change
   adaptation, v, 2, 3, 6–10, 16, 21,
      26–28, 41, 43, 46–49, 71, 72,
      75, 76, 88, 92–94
   blue carbon, vi, 10, 33, 42, 88–91
   carbon capture and storage (CCS),
      42, 46, 47, 58, 65, 87–90, 94
   carbon sinks, vi, 16, 33, 42, 65,
      87–91
   flooding, 1, 3, 4, 6, 9, 16, 17, 30,
      33, 42, 56, 73, 74, 77, 78, 81,
      82, 89, 92, 94
   mitigation, v, 2, 4, 6–9, 16, 21, 27,
      28, 43, 46, 48–50, 71, 72, 87,
      88, 93, 94
   sea level rise, 1, 4, 6, 9, 16, 17, 30,
      33, 46, 56, 72, 77, 81, 82, 90,
      92–94
   tipping points, 17, 18, 49, 72
   wetlands, vi, 8, 16, 40–43, 46, 49,
      73, 77, 81, 87–89

D

Davidson, J., 40
Davidson-Hunt, J., 34
   co-production of knowledge, 10,
      28, 34, 72

E

Environmental Protection Agency
      (EPA)
   Exclusive Economic Zone and
      Continental Shelf
      (Environmental Effects) Act,
      2013 (EEZ), 8
   green house gas (GHG)
      emissions, 7–9, 15, 16, 21,
      21n1, 42, 55–64, 66, 67, 76,
      87, 88, 90, 93

He Whetū Mārama, 47, 48, 50
Ministry for the Environment
      (MfE), 58, 76
natural hazard management, 9, 72
New Zealand Emissions Trading
      Scheme (NZETS), 8, 9, 21,
      43, 47, 50, 56, 58–67, 76,
      77, 87, 88
Resource Management Act, 1991
      (RMA), 8, 43–46

H

Hutchings, J., 45
Iwi Management Plans (IMP), 45

I

Indigenous knowledge
   experiential knowledge, 5, 20, 29,
      40, 93
   ki uta ki tai, 47, 49
   place names, 5, 40
International Panel on Climate
      Change (IPCC), 1, 2, 4, 8, 17,
      21, 27, 28, 56, 77, 91, 93

K

Kāi Tahu
   Kāti Huirapa Rūnaka ki Puketeraki
      (KHR), 48, 52n39
   Rūnaka, 48, 49, 52n39, 52n41, 75
   Strategy and Influence team, 49, 50,
      52n40, 52n41
   Te Rūnaka o Ōtākou, 48, 50,
      52n34, 52n39
   Te Rūnanga o Ngāi Tahu (TRoNT),
      49, 52n39, 52n40, 52n41
   Te Wai Pounamu, 32, 37n38,
      49, 89
Kyoto Protocol

carbon credits, 59
Climate Change Response Act,
    2002, 58
Climate Change Response
    (Emissions Trading)
    Amendment Act, 2008, 58
Climate Change Response
    (Moderated Emissions Trading)
    Act, 2009, 60
Cognitas Report, 63, 64
Comprehensive Strategy on climate
    Change (CSCC), 57
National Communication
    (1994), 58

L
Land
    cultural landscapes, 8, 18, 64
    deforestation, 16
    kaitiaki, 10, 64, 72
    land-sea interface, 10
    land-use changes, 8, 39–50, 58,
        65, 77
    mahika kai, vi, 18, 40, 81
    takiwā, 10, 49, 72

M
Māori
    ahi kā, 20
    hapū, 62
    iwi/tribes, 3, 19, 21, 26, 44, 56,
        62, 63, 72, 74
    tūrakawaewae, 20
    whānau, 62, 65, 77
Mātauraka-a-Iwi, 10, 26, 49, 88
    Māori Environmental Knowledge
        (MEK), 6, 8, 10, 17, 19–21,
        26, 33, 43, 44, 47, 48, 50, 63,
        75, 76, 79, 86, 91, 93, 94

Mātauraka Māori (MM)
    ira atua, 20
    ira takata, 20
    kaitiakitaka, 20, 44
    mana whenua, 47, 81
    rāhui, 17
    taonga tuku iho, 64
    tikaka, 21
    whakawhanaukataka, 34, 35, 65

N
National Institute of Water and
    Atmospheric Research (NIWA),
    27, 49, 50, 74, 75
    Arowhenua Pā, 72–75
Nunn, P. D., 31, 32

P
Pacific Island Countries (PIC), 1, 3, 4,
    6, 7, 10, 16, 18, 28, 34, 55, 56,
    87, 91–94
Park, G., 41
Polynesian Triangle, 6, 11

T
Traditional ecological knowledge
    (TEK), v–vii, 7, 10, 25–35, 73,
    81, 85, 86, 91–94
    knowledge framework, 6, 8, 25, 28
Tranter, B., 2
Treaty of Waitangi
    Crown, 62
    Local Government New Zealand
        (LGNZ), 44
    New Zealand Government, 62
    partnership, protection,
        participation, potential, 67
    treaty partners, 44, 62

U
United Nations (UN)
 United Nations Development
  Programme (UNDP), 27
 United Nations Educational,
  Scientific and Cultural
  Organisation (UNESCO),
  27, 90
United Nations Food and
 Agriculture Organisation
 (UNFAO), 92
United Nations Framework
 Convention on Climate
 Change (UNFCCC), 3,
 57, 58
United Nations University, 27